Instructor's Resource Manual
to accompany

Physics

A Practical and Conceptual Approach

SECOND EDITION

JERRY D. WILSON

Chairman, Department of Science and Mathematics
Lander College

Saunders Golden Sunburst Series

SAUNDERS COLLEGE PUBLISHING

Philadelphia
Fort Worth
Chicago
San Francisco
Montreal
Toronto
London
Sydney
Tokyo

Wilson: Instructor's Resource Manual to Accompany
        PHYSICS: A PRACTICAL AND CONCEPTUAL APPROACH, Second Edition

ISBN #   0-03-023772-6

901   018   987654321

Preface

This Resource Manual was prepared to supplement **Physics: A Practical and Conceptual Approach.** It provides the instructor with the following aids:

For each chapter

* <u>Answers to Questions</u>. General answers are given for the Questions at the end of each chapter. The instructor may wish to elaborate on these in many cases.

* <u>Sample Test Questions</u> (Test Bank). There are in general 25 or more of each of the following type questions (with answers):

    a) Multiple Choice

    b) Completion

    c) Matching

    In total, over 2000 questions form a Test Bank to assist in the preparation of quizzes and tests through the selection of various types of questions on the material specific for a particular class. The Test Bank is also available on diskette for your convenience should you prefer to prepare tests by computer.

Given at the end of the manual are

* <u>Solutions to Extended View Exercises</u>. Worked-out solutions to the simple optional exercises. These exercises have been added at the end of the text for some chapters for the instructor who believes some simple mathematical analysis is of instructional benefit in his or her course.

It is hoped that you will find these features helpful.

Thanks is given to James T. Shipman who organized this manual for the second edition of the text and prepared many sample test questions and to Ruth Hodges who typed the Resource Manual.

Jerry D. Wilson

Lander College

Chapter 1   A Restless World: Motion, Force, and Newton's Laws

Answers to Questions

1. Change the position of furniture and other household items.

2. Opportunity to change position of a checker.

3. Yes, with finish point same as starting point.

4. Approximate instantaneous speed.

5. Instantaneous speeds.

6. Zero.

7. Yes. A negative acceleration.

8. An acceleration is produced when there is a change in the velocity of the car. Since velocity is a vector, an increase or decrease in magnitude and/or a change in direction will produce an acceleration. A steering wheel is an accelerator as a "direction changer". Gearing down is a decelerating action.

9. Yes. Direction, and hence velocity, is changing.

10. (a) At center. (b) Towards back.

11. Acceleration is evidence of a force, seen or unseen.

12. Balloon goes forward (similar to air bubble level).

13. There is an equal or greater opposing force.

14. Actual food weight exclusive of packaging.

15. Yes. Constant velocity.

16. Galileo correctly considered that an object exhibited the behavior of maintaining a state of motion rather than that the normal state of a body was at rest (Aristotle).

17. No. It is a general property of matter.

18. Inertia of the plates and glasses resists changes in motion.

19. Inertia of roll resists sudden change in motion of a large force. Works better for large roll (more inertia).

20. The inertia of the head.

21. To supply force so person doesn't continue moving.

22. (a) and (b) "thrown" backwards due to inertia, (c) stands easily, (d) goes forward, (e) goes away from corner or "thrown outward", but actually continues according to Newton's first law.

23. (a) $ α n, (b) No, (c) $ = 0.05n, (d) Yes, for n = 10, $ = 0.05(10) = 0.50, or $0.50.

24. Inverse proportion, study time α 1/extracurricular activities.

25. (a) 2F/m = 2a, (b) F/m/2 = 2a, (c) 2F/2m = a

26. Push them with equal force. Heavier object would accelerate less (more inertia).

27. Push or lift the containers. Full container would accelerate less (more inertia).

28. If F = 0, a = 0 and velocity is constant.

29. (a) Net force is zero, acceleration is zero.
    (b) Equal and opposite forces transmitted by rope.

30. No. A constant acceleration indicates the falling body is gaining velocity at a constant rate. Since the velocity is increasing, the distance traveled per unit time will increase.

31. The diver in the fetal position because this diver will fall faster due to less air resistance.

32. Because of the opposing force produced by the updraft, the terminal velocity of a falling object will be decreased.

33. Sleet is smaller in size and area than snow flakes, therefore the air resistance will slow the snow flakes quicker, reaching therminal velocity first.

34. Yes. When falling in a vacuum where there is no opposing forces such as air resistance, or in air when a lighter object is more streamlined than a heavier object.

35. She will accelerate to a greater terminal velocity.

36. No. The moon has no atmosphere.

37. Take off skates and help.

38. (a) Spring force on person, (b) Reaction force of wall on swimmer, (c) Reaction force of canoe. Both are accelerated.

39. Hand force and wall reaction force equal and opposite, so no net force.

40. (a) Vertical action forces of tires on road and reaction forces of road on tires. (b) Horizontal reaction forces on the road surface and equal and opposite forces of static friction on the tires in the direction of motion that accelerates the car. Also, vertical force pairs as in (a).

41. (a) Reaction force to force on water.
(b) Upward reaction force of air.

42. (a) 100 N, (b) 100 N

43. (a) 9.8 N, (b) It would overlap itself. When connecting rope is moving, as in pulley system, there is a net tension between ends that accelerates rope as part of system. In this case, force is transmitted undiminished in the approximation rope has zero mass.

44. Reaction force on spacecraft is opposite to direction of motion.

45. (a) Equal and opposite to the upward action force of raising the arms is a downward reaction force, which causes the scales to read heavier. (b) Reverse situation, scales read lighter.

SAMPLE TEST QUESTIONS

Multiple Choice

1. A change in motion (a) requires a balanced force, *(b) is evidence of an unbalanced force, (c) is produced only by contact forces, (d) can occur if the acceleration is zero.

2. The property of inertia (a) is the same as friction, (b) assists changes in motion, (c) applies only to very massive bodies, *(d) is related to the mass of a body.

3. Which of the following is <u>not</u> a vector quantity? *(a) speed, (b) velocity, (c) acceleration, (d) force

4. An acceleration can result from (a) a change in speed, (b) a change in direction, (c) changes in both speed and direction, *(d) all of the preceding.

5. If two equal and opposite forces act on a body, the body will (a) experience and acceleration, (b) increase its inertia, *(c) remain at rest or in uniform motion, (d) experience an unbalanced force.

6. When the net force acting on a body increases, (a) its velocity remains uniform, (b) its acceleration remains constant, *(c) a change in speed and/or direction occurs, (d) its speed always increases.

7. Which of the following is a unit of acceleration?

   *(a) m/s², (b) km-cm/s, (c) s-s, (d) cm/s

8. Friction occurs (a) only between solid surfaces, *(b) between all contacting media (solid, liquid, and gas), (c) primarily due to interlocking surface irregularities, (d) when an object moves in a vacuum.

9. Weight (a) is always less in magnitude than an object's mass, (b) has the units of kilos, *(c) is a gravitational force, (d) is the same as mass.

10. The action and reaction forces of Newton's third law (a) is the same as the unbalanced force F in Newton's second law, (b) are in the same direction, (c) are always contact forces, *(d) act on different bodies.

11. Motion is a process of a change in (a) time, (b) mass, (c) weight, *(d) position.

12. The laws of motion were formulated by (a) Aristotle, (b) Galileo, *(c) Newton, (d) Einstein.

13. Which of the following is a scalar quantity? *(a) speed, (b) velocity, (c) acceleration, (d) force.

14. In addition to a magnitude, a vector quantity has (a) length, (b) time, *(c) direction, (d) an average value.

15. The time rate of change of velocity is (a) speed, *(b) acceleration, (c) friction, (d) force.

16. Two equal and opposite forces of 3 N have a net force of (a) 9 N, (b) 6 N, (c) 3 N, *(d) 0 N.

17. The acceleration of an object acted upon by a net force perpendicular to the direction of the object's original motion is (a) a deceleration, (b) zero, (c) not covered by Newton's laws, *(d) due to a change in direction.

18. When the force of static friction opposes an applied force, (a) there is an acceleration, (b) the acceleration due to gravity changes, *(c) the net force is zero, (d) there is no motion due to inertia.

19. Weight and mass (a) are the same thing, *(b) differ in magnitude by a constant, (c) are a result of friction, (d) are defined by the first law of motion.

20. The action and reaction of Newton's third law *(a) act on different objects, (b) are always due to gravity, (c) are expressed in the law of inertia, (d) are accelerations.

21. All objects in free fall near the Earth's surface have the same (a) velocity (b) speed, *(b) acceleration, (c) weight.   d all   e a + b only

22. When an object falls toward the Earth with only _____ acting on it, we say it is in free fall. (a) friction, *(b) gravity, (c) downward forces, (d) vertical forces.

23. The air resistance on a falling body depends on its shape, exposed area, and (a) composition, (b) mass, *(c) speed, (d) none of these.

24. An automobile is traveling due east on an interstate highway at a constant velocity of 65 miles per hour. The unbalanced force acting on the car with respect to the highway is (a) toward the east, (b) toward the west, (c) directed vertically downward, *(d) zero.

25. Mass is a measurement of (a) volume, (b) weight, *(c) inertia, (d) none of these.

26. Speed (a) is a scalar quantity, (b) has the units of length over time, (c) has no direction, *(d) all of these.

27. Two objects of different mass have the same size and shape. When they are dropped from a height of 100 meters above the Earth's surface, (a) the less massive object will reach the ground first, (b) the more massive object will reach the ground first,*(c) they will reach the ground at the same time, (d) none of the above.

28. Newton's third law of motion states that for every force there is an equal and opposite force and the forces (a) are on the same object, *(b) are on different objects, (c) must be under the influence of gravity, (d) produce an acceleration.

29. Newton's third law of motion states that the forces acting between two different masses are (a) equal and in the same direction, *(b) equal and in the opposite direction, (c) unequal and parallel, (d) equal and perpendicular.

30. Velocity (a) has no direction, (b) has the units of displacement, (c) is a scalar quantity, *(d) is a vector quantity.

31. A projectile *(a) has a constant speed in the horizontal direction, (b) is always projected in one dimension, (c) has no forces acting on it, (d) has no vertical acceleration.

32. An object in free fall on Earth has (a) an acceleration that depends on its mass, (b) air resistance acting on it, (c) a constant velocity, *(d) an acceleration of 9.8 m/s^2

33. An object in nonfree fall with air resistance (a) falls with a constant acceleration, (b) has an increase in velocity of 9.8 m/s each second, *(c) has a continually diminishing acceleration, (d) may accelerate beyond its terminal velocity.

34. A sky diver falling at terminal velocity (a) is in free fall, *(b) has a net force of zero, (c) has an acceleration due to gravity, (d) must fall from rest.

35. For which projection is the horizontal acceleration zero? (a) vertical projection, (b) horizontal projection, (c) projection at an angle, *(d) all of the preceding.

36. The type of path an object projected at an angle (other than 90°) follows *(a) is a parabolic arc, (b) is a circular arc, (c) depends on its initial velocity, (d) depends on the angle of projection.

37. For an object in free fall, (a) the momentum is conserved, (b) there is zero net force acting on it, (c) equal distances are travelled in equal times, *(d) only the gravitational force is considered.

38. Galileo's legendary Leaning Tower of Pisa experiment, (a) showed objects of different weights fall at different rates, (b) confirmed Aristotle's views on motion, (c) was actually done in Venice, *(d) is seriously questioned with regard to authenticity.

    only grav. acting

39. For two objects of different mass in free fall, (a) the accelerations are different, *(b) the acting forces are different, (c) air resistance is a consideration, (d) the more massive object will reach the ground first if released simultaneously.

Completion

1. Newton published the laws of inertia in his famous book called "Principia".

2. Motion is the process of a change in position.

3. Instantaneous speed is the speed of an object at a particular time.

4. A change in velocity is given by the product of acceleration and time.

5. The net force of equal and opposite forces is zero.

6. Force is a vector quantity.

7. Inertia is the property of mass that resists changes in motion.

8. The acceleration of an object is inversely proportional to its mass.

9. The acceleration of an object with a zero net force acting on it is zero.

10. Jet propulsion of rockets is a good example of the third law.

11. Acceleration is the time rate of change of velocity.

12. Forces that cancel each other are called <u>balanced forces</u>.

13. The <u>newton</u> is the SI unit of force.

14. The force of friction resists <u>motion</u>.

15. Force is a quantity <u>capable</u> of producing motion or a change in motion.

16. A change in motion is indicative of a <u>force</u>.

17. The greater the <u>mass</u> of an object, the greater its inertia.

18. <u>Length</u> and <u>time</u> are the fundamental properties used to describe motion.

19. <u>Velocity</u> specifies speed and direction.

20. For every action, there is an equal and opposite <u>reaction</u>.

21. Distance traveled per unit time is called <u>speed</u>.

22. Displacement divided by time is called <u>velocity</u>.

23. A continuous change in position is called <u>motion</u>.

24. The straight-line distance between two points is called <u>displacement</u>.

25. The force of attraction between a mass and the Earth is called <u>weight</u>.

26. The weight of an object can be calculated by multiplying the mass of the object by <u>gravity</u>.

27. A <u>scalar</u> quantity has magnitude only.

28. Newton's <u>first</u> law of motion refers to masses at rest or in motion.

29. When an object falls toward the Earth with only gravity acting on it, we say the object is in <u>free fall</u>.

30. All objects in free fall near the surface of the Earth fall with the same <u>acceleration</u>.

31. The air resistance on a falling object is a function of the object's size, exposed area, and <u>speed</u>.

32. When an object falling through the Earth's atmosphere no longer accelerates but falls with a constant speed, we say that the object has reached <u>terminal</u> velocity.

33. Any quantity capable of producing motion is called a <u>force</u>.

34. A net or resultant force always produces an <u>acceleration</u>.

35. <u>Mass</u> is a measure of inertia.

36. The metric SI unit of force is the <u>newton</u>.

37. When an object falls toward the Earth with only gravity acting on it, we say it is in <u>free fall</u>.

38. The air resistance on a falling body depends on its size, speed, and <u>shape</u>.

39. A falling object reaches <u>terminal</u> velocity when the force of air resistance equals the weight of the falling object.

40. Two metal balls having the same size and shape but with different mass are dropped an altitude of 2 km above the Earth's surface. The <u>more</u> massive ball reaches the ground first.

41. According to Newton's <u>first</u> law of motion, a body at rest will remain at rest unless acted upon by an unbalanced force.

42. The time rate of change of position is known as <u>speed</u>.

43. Velocity is a <u>vector</u> quantity.

44. The amount of force that gives a mass of one kilogram an acceleration of 1 m/s² is one <u>newton</u>.

45. Kinetic friction is generally <u>less</u> than static friction.

46. According to Newton's <u>third</u> law of motion, forces always occur in pairs.

47. A net force acting on a body produces an <u>acceleration</u>.

48. Speed is a <u>scalar</u> quantity.

49. A vector quantity has magnitude and <u>direction</u>.

50. Galileo originated the law of <u>inertia</u>.

51. If the net force acting on an object is doubled, the acceleration is <u>doubled</u>.

52. Friction always opposes relative <u>motion</u>.

Chapter 1

Matching

(Choose the appropriate answer from the list on the right.)

__d__ 1. Capable of producing motion   a. newton (unit)

__f__ 2. rate of change of position   b. second law of motion*

__k__ 3. units of acceleration   c. rocket*

__a__ 4. unit of force   d. force

__r__ 5. magnitude and direction   e. third law of motion

__n__ 6. laws of motion   f. speed

__j__ 7. law of inertia   g. inertia

__m__ 8. acceleration of an object during free fall   h. resultant

__h__ 9. vector sum   i. friction

__i__ 10. static and kinetic   j. first law of motion

__g__ 11. measure of mass   k. m/s²

__l__ 12. zero acceleration   l. constant velocity

__e__ 13. action and reaction   m. independent of mass

n. Newton

o. m/s*

p. Aristotle*

q. deceleration*

r. vector

*Answers not used

12

Matching

(Choose the appropriate answer from the list on the right.)

<u>d</u> 1. weight

<u>p</u> 2. motion

<u>k</u> 3. net force

<u>g</u> 4. Newton's first law

<u>l</u> 5. inertia

<u>o</u> 6. speed

<u>m</u> 7. free fall

<u>b</u> 8. velocity

<u>q</u> 9. acceleration

<u>j</u> 10. Newton's second law

<u>i</u> 11. terminal velocity

<u>f</u> 12. friction

<u>a</u> 13. Newton's third law

a. action and reaction

b. speed and direction

c. meter*

d. mass x gravity

e. balanced force*

f. resists motion of contacting surfaces

g. law of inertia

h. always zero*

i. function of air resistance

j. F = ma

k. produces an acceleration

l. resists changes in motion

m. a = 9.8 m/s^2

n. negative acceleration*

o. distance/time

p. changing position

q. change in velocity/time

*Answers not used

Chapter 2   Work and Energy

Answers to Questions

1. Yes, in the sense that we can feel the reception of
    energy, for example, the reception of sunlight.
    Other senses are excited by the expenditure of
    energy.

2. Energy is defined as the ability to do work. Energy
    is a concept. A concept is defined as a meaningful
    idea that can be used to describe phenomena.

3. No, not doing work. Yes, has done work in lifting
    the weights (force x distance).

4. (a) The tall weight lifter.
    (b) Have both lifted equal weights through the same
    distance.

5. Yes, force and displacement (vectors) in opposite
    directions. The force acts opposite the motion,
    tending to slow the object down.

6. (a) If you and the student weigh the same, then you
    both do the same amount of work. If you are of
    different weights, then the heavier person will
    do more work.
    (b) You do (assuming equal weights). Power equals
    time rate of doing work.

7. The weight (force) of the bucket of water times the
    depth (distance) of the well (plus any frictional
    losses).

8. (a) Friction   (b) To reduce friction

9. Yes, but the smaller h.p. motor would do it slower.
    The larger motor could do the work three times as
    fast.

10. Piece work takes into account energy/time (power).
    Hourly rates hold the time period effort fixed and
    the power output is arbitrary since the incentive
    (money) is constant and there is usually variable
    work output.

11. More power is required to plow.

12. (a) The power outputs are not the same, since
    greater acceleration with 2F.
    (b) For $P_2 = 2P_1$, requires $t_2 = t_1$. Since $t_2 < t_1$,
    P_2 is more than twice as large as P_1.

13. To promote water flow to potential energy resulting from the height of the tower.

14. Due to elasticity. When an object is shot, the potential energy is transformed into kinetic energy.

15. $\frac{1}{2}mv^2 = mgh$, and $v = (2gh)^{\frac{1}{2}}$

16. Yes, Reference of zero potential energy is arbitrary.

17. Yes, due to motion of car and relative to the road.

18. If the work is positive, an increase in kinetic energy, and if negative, a decrease in KE.

19. Lifting an object, negative work is "stored" energy. Falling object, positive work, decrease of potential energy, increase in kinetic energy. Energy is used to do work when hitting ground if "dent" in ground or floor is made and dissipated as heat and sound.

20. More mass or inertia, and more kinetic energy or momentum transfer.

21. Exchange of internal energy, equal and opposite forces. Maximum height determined by elasticity and energy exchange.

22. (a) Potential energy exchange into work.
    (b) Greater distance of fall with each driven distance, more energy for greater driving distance.

23. An amount equal to its kinetic energy.

24. (a) $mv_2 = m(3v) = 3mv$ (three times).
    (b) $\frac{1}{2}m\tilde{v}^2 = \frac{1}{2}m(3v)^2 = 9(\frac{1}{2}mv^2)$ or nine times.

25. Assuming same braking force, car with 4 times braking distance does 4 times as much work and loses 4 times as much energy, going half the speed ($W \propto v^2$).

26. Goes into useful work and lost energy (for example, heat).

27. Another chemical compound is formed and useful energy released.

28. Involves a force or a "disturbance" acting through a distance.

29. Mechanical, electrical, heat, chemical, nuclear

30. Not in the classical sense. Matter is converted into energy and vice versa in nuclear processes. But in the sense that mass is a form of energy, it is not created or destroyed.

31. (a) conversion of chemical energy to electrical energy, (b) conversion of the energy of the wind (kinetic) to energy of the sailboat (kinetic) (c) conversion of solar energy to electrical energy, (d) conversion of potential energy (position of water) to electrical energy (Answer for part b of question) The Sun.

32. Goes into work against friction (dissipated as heat) in stopping the car.

33. Converted to potential energy (and loss due to friction).

34. Converted to frictional losses and work of digging up dirt.

35. 50 J

36. Gravity on moon is less. Less work (mgh), slower speed, but same height (conservation of energy).

37. Energy expended and work done against friction (converted to heat).

38. Energy (kinetic) converted to work (heat) against friction.
Depends on distance and/or (frictional) force of skid.

39. (a) mgh. (b) Yes, the initial potential energy must account for both kinetic and potential energies at the top of the loop.

40. (a) The mechanical advantage is the ratio of the length of the lever arms. This relates to the ratio of the distance the applied force moves to the distance the load moves. (b) No mechanical advantage.

41. $F_o L_o = F_i L_i$ with no friction, and $F_o/F_i = L_i/L_o$

42. (a) To have a single person go a greater height. (b) Yes. Greater weight, greater potential energy or work required. (c) Two times the force and 3 times the lever arm, the work output is 6 times greater and the person goes 6 times higher.

43. Five supporting strands.

44. Less upward force required at the expense of distance. Work done against only a component of gravity.

45. Efficiency is defined as work out (energy) divided by work in (energy). The more efficient the machine the more useful the output will be and less energy will be wasted.

46. (a)   No.   (b) The inputs would have to be twice their rated power outputs.

47. E_1 = 500/700 = 5/7 (= 10/14), E_2 = 900/1400 = 9/14.

48. The energy efficiency is very low. Water is heated unnecessarily.

SAMPLE TEST QUESTIONS

Multiple Choice

1. Energy is (a) a tangible thing, *(b) something abstract, (c) not associated with fundamental forces, (d) a vector quantity.

2. Work is done by a Force (a) by any applied force, (b) by all components of force, *(c) when a force moves an object, (d) by an applied force perpendicular to the direction of motion.

3. A unit of work or energy is the (a) N/m, (b) ft-s, (c) kg-m, *(d) kg-m^2/s^2

4. Power (a) is work times time, (b) has units of joules, (c) is the same for all motors, *(d) is the time rate of doing work.

5. The gravitational potential energy (a) is independent of height, (b) is always positive, *(c) is independent of path, (d) decreases with increasing height.

6. Kinetic energy is (a) the energy of position, *(b) always positive, (c) independent of mass, (d) a vector quantity.

7. The form of energy in which mass conversion is significant is (a) electrical, (b) gravitational, *(c) nuclear, (d) chemical.

17

Chapter 2

According to the 1st Law of Thermo.

8. The total energy is conserved in (a) a conservative system, (b) a nonconservative system, (c) the universe, *(d) all of the preceding.

9. A machine *(a) can have a mechanical advantage greater than one, (b) multiplies the work input, (c) can run perpetually, (d) is not subject to the conservation of energy.

10. The efficiency of a machine (a) can be greater than one, (b) has the units of joules, (c) is the force multiplication factor, *(d) expresses the fraction of useful work done.

11. Work requires (a) a zero net force, (b) a force perpendicular to the direction of motion, *(c) motion, (d) zero momentum.

12. The unit of work is *(a) joule, (b) m/s^2, (c) lb, (d) N.

13. The time rate of doing work is (a) energy, *(b) power, (c) momentum, (d) efficiency.

14. A form of mechanical energy is (a) electromagnetic, (b) nuclear, (c) efficiency, *(d) kinetic.

15. The N-m/s is a unit of (a) work, (b) energy, *(c) power, (d) efficiency.

16. The energy of position is (a) heat, (b) work, (c) kinetic energy, *(d) potential energy.

17. Energy is (a) a vector quantity, *(b) the ability to do work, (c) always used in useful work, (d) has only two general forms.

18. The energy of a nonconservative system (a) is all mechanical energy, (b) may all be converted to useful work, *(c) is totally conserved, (d) has 100 percent efficiency when converted to work.

19. A machine is used to (a) decrease efficiency, (b) increase energy, (c) increase work output, *(d) change the magnitude of a force.

20. When the efficiency of a machine is increased, *(a) the work output increases, (b) the work input is increased, (c) perpetual motion occurs, (d) the total energy is not conserved.

21. The watt is a unit of (a) energy, (b) work, *(c) power, (d) none of these.

22. One horsepower is equivalent to (a) 550 ft-lb/min, (b) 476 watts, (c) 555 ft-lb/s, *(d) 746 watts.

23. Which of the following signifies the greatest amount of power? (a) 1 watt, *(b) 1 horsepower, (c) 1 ft-lb/s, (d) 1 joule/s.

24. A kilowatt-hour is a unit for (a) electricity, *(b) energy, (c) power, (d) all of these.

25. When the velocity of a mass is doubled, the kinetic energy (a) is doubled, (b) is halved, *(c) is quadrupled, (d) remains the same.

26. The kinetic energy of a moving mass is proportional to the (a) velocity, *(b) square of the velocity, (c) square root of the velocity, (d) none of these.

27. The mechanical advantage of a machine can be calculated by dividing the (a) work in by the work out, (b) work out by the work in, (c) applied force by the load, *(d) load by the applied force.

28. One watt is equal to _____ per second. (a) one newton, *(b) one joule, (c) 550 ft-lb, (d) one meter.

29. One horsepower is equal to _____ ft-lb/s. (a) 746, (b) 33,000, *(c) 550, (d) none of these.

30. Which of the following is not a unit of energy? (a) foot-pound, (b) joule, (c) kilowatt-hour, *(d) horsepower.

31. Work is equal to (a) force x any distance, *(b) power x time, (c) energy x time, (d) force x mass.

32. The magnitude for kinetic energy is equal to (a) mv/2, (b) 2 mv², (c) m²v/2, *(d) m²v/2

33. The work done by a force acting on an object is defined as the product of the force and the distance through which the object moves ____ to the force. (a) perpendicular, *(b) parallel, (c) in any direction, (d) all of these.

34. Which of the following is not a form of energy? (a) heat, (b) light, *(c) joule, (d) electricity.

19

35. The _____ multiplication factor is called the mechanical advantage of a machine. (a) power, (b) work, (c) energy, *(d) force.

36. One person can do a certain amount of work in a given time. A second person does the same work in half that time. The power developed by the second person relative to the first is _____ as much. (a) one-half, *(b) twice, (c) four times, (d) one-fourth.

Completion

1. A net force and/or *motion* must be present in order to have work. *(movement)*

2. A force acting through a distance results in *work* being done.

3. The SI unit of work is the *newton-meter or joule*.

4. Work divided by the amount of time required to do the work is called *power*.

5. The two types of mechanical energy are *potential, energy*, and *kinetic energy*.

6. Gravitation potential energy is the product of *weight* and *height*.

7. *Kinetic energy* is equal to one-half times the product of the mass and the square of the velocity.

8. *Energy* is commonly defined as the ability to do work.

9. The random kinetic energy of the atoms in matter is called *heat energy*.

10. According to the law of conservation of energy, energy cannot be *created* or *destroyed*.

11. Work is a *scalar* quantity.

12. *Power* is the time rate of change of doing work.

13. A body has *energy* when it is capable of doing work.

14. The kinetic energy of a body is proportional to the *square* of its velocity.

15. The total energy is *always* conserved.

16. The location of h = 0 for gravitational potential energy is <u>arbitrary</u>.

17. The work done in stopping a moving object on a horizontal surface is equal to the object's <u>kinetic energy</u>.

18. Mechanical energy is conserved in a <u>conservative</u> (ideal) system.

19. Pulling a nail with a claw hammer is an application of a simple machine called the <u>lever</u>.

20. When the work "lost" increases in a machine, its efficiency <u>decreases</u>.

21. Work is defined as <u>force</u> times parallel distance.

22. The unit for work in the British system is the <u>foot-pound</u>.

23. Energy is defined as the ability to do <u>work</u>.

24. The SI unit for power is the <u>watt</u>.

25. The British system unit for power is the <u>horsepower</u>.

26. Any device used to change the magnitude or direction of a force is called a <u>machine</u>.

27. The force multiplication factor of a machine is known as <u>mechanical advantage</u>.

28. The ratio of the work done by a machine to the energy supplied to it is called its <u>efficiency</u>.

29. A kilowatt-hour is a unit of <u>energy</u>.

30. One watt is one <u>joule</u> per second.

31. A mass that is moving will always have <u>mechanical</u> <u>kinetic</u> energy.

32. Potential energy is the energy of <u>position</u>.

33. A component of force must be <u>parallel</u> to the direction of motion in order for work to be done.

34. A 1-hp motor does <u>one-third</u> the amount of work done by a 3-hp motor in the same amount of time.

35. A mechanical advantage can increase the force but at the expense of <u>distance</u>.

36. A single fixed pully has a mechanical advantage of <u>one</u>.

37. The <u>efficiency</u> of a machine is the ratio of the output divided by the input.

Chapter 2

Matching

(Choose the appropriate answer from the list on the right.)

_h_ 1. work   a. joule per second

_l_ 2. joule   b. work/time

_b_ 3. power   c. work out/work in

_e_ 4. horsepower   d. proportional to velocity squared

_o_ 5. energy of motion e. 746 watts

_j_ 6. mechanical energy f. motion of electric charge

_f_ 7. electrical energy g. 100% efficiency

_m_ 8. nonconservative system h. force x distance

_c_ 9. efficiency   i. mgh*

_g_ 10. perpetual motion machine j. KE + PE

_a_ 11. watt   k. light

_k_ 12. a form of energy l. newton-meter

_d_ 13. kinetic energy   m. mechanical energy lost

                         n. mechanical advantage*

                         o. kinetic energy

                         p. N/s*

                         q. KE - PE*

*Answers not used.

23

Chapter 2

Matching

(Choose the appropriate answer from the list on the right.)

_f_	1.	work	a.	ability to do work
_a_	2.	energy	b.	watt
_i_	3.	power	c.	100% efficiency
_l_	4.	potential energy	d.	conservative system
_o_	5.	kinetic energy	e.	nuclear
_d_	6.	conservation of mechanical energy	f.	force x distance
_g_	7.	machine	g.	force multiplier
_k_	8.	mechanical advantage	h.	work out/work in
_h_	9.	efficiency	i.	work/time
_c_	10.	perpetual motion machine	j.	joule
_b_	11.	unit of power	k.	force multiplication factor
_j_	12.	unit of energy	l.	mgh
_e_	13.	a form of energy	m.	energy saver*
			n.	total energy*
			o.	mv²/2
			p.	F/d*
			q.	always independent of mass*

*Answers not used.

Chapter 3   Momentum

Answers to Questions

1. The padded mats increase the impulse time, thereby decreasing the impulse or impact force on landing.

2. Greater contact time, less impulse force for a given momentum transfer (collision).

3. No. Greater contact time, less impulse force for a given momentum transfer (collision).

4. Small contact time and force so ball doesn't receive large momentum and go too far. Some of force is usually dissipated in sand.

5. Impulse in one direction gives impulse in opposite direction (conservation of momentum), for example, a pulse jet engine.

6. Greater contact time, less forceful blow.

7. Increase contact time decreases impulse force. Similar to moving hands backward when catching a ball.

8. Short jabs, quick contact time, but not much force, just "stinging" blows. Follow through gives greater impulse, compounding the forceful blow.

9. Greater contact time, less impact force.

10. Conservation of momentum. Air goes out back, balloon goes forward, but in random direction.

11. Head-on collision gives greater momentum transfer and greater damage.

12. More powder used to overcome greater inertia of more massive bullet and more momentum transfer. No, you might shoot yourself.

13. The equal and opposite forces are on the projectile and escaping gases, not on launcher. It is dangerous behind because of escaping hot gases which carry large momentum.

14. Some of gas deflected forward to reduce recoil.

15. By conservation of momentum, hose moves backwards and fire fighter must apply forward force.

16. It moves in the opposite direction, but because of the hugh mass of Earth, the motion is negligible.

17. Changing direction of exhaust gases gives opposite reaction in desired direction.

18. Exhaust gases directed toward moon. By conservation of momentum (or Newton's 3rd law), force on rocket is away from moon.

19. No, momentum changes because of a change in direction (vector quantity), and $\Delta p = p_2 - p_1 = mv - (-mv) = 2 mv$.

20. The astronaut has to apply a force to another mass in a direction opposite the spaceship in order to apply a force on his or her body in the direction of the spaceship. (Newton's third law of motion) The astronaut can throw a tool or some other mass off into space.

21. Internal force, and by conservation of momentum, no change in momentum. With fan reversed, similar to rocket.

22. (a) Yes. (b) No, there is an acceleration (due to gravity) in this direction.

23. Zero before, zero after (no external force).

24. If momentum is conserved for collision of equal masses, so is velocity (masses cancel). For unequal masses, different velocities, so not conserved, but momentum is.

25. m, complete momentum transfer.

26. (a) After collision, the lighter incoming ball will rebound in the opposite direction. The heavy stationary ball will move slightly in the original direction of the lighter ball. (b) After collision, the heavy ball will continue in the same direction with slightly less velocity. The light stationary ball having received some momentum from the heavier ball moves in the same direction as the heavier ball but with a greater velocity than the heavier ball.

27. Conservation of horizontal momentum.

28. (a) Yes. Colliding objects are elastically
    deformed, so mechanical energy is conserved.
    (b) No. Only some of the kinetic energy is lost.
    Momentum would not be conserved with complete loss
    (v = 0).

29. At 45° or NE.

30. They are stationary.

SAMPLE TEST QUESTIONS

Multiple Choice

1. Momentum is (a) a vector quantity, (b) dependent on
    mass, (c) a function of velocity, *(d) all of these

2. Momentum has the units of (a) newton x meter, *(b)
    newton x second, (c) newton/meter, (d)
    newton/second.

3. The units of momentum are (a) force x velocity, (b)
    mass x acceleration, (c) different from impulse,
    *(d) mass x velocity

4. Momentum (a) is a scalar quantity, (b) is
    independent of mass, (c) is always constant, *(d)
    has the same units as impulse.

5. Momentum takes into account (a) space and time, (b)
    collisions and heat, *(c) inertia and motion, (d)
    shape and size.

6. Momentum is conserved (a) in an elastic collision of
    two balls, (b) in an inelastic collision of two
    balls, (c) in the absence of an external unbalanced
    force, *(d) in all of the preceding cases.

7. The momentum of a system is conserved when (a) the
    net internal force is zero, (b) the external net
    force is nonzero, (c) the impules on the system is
    increased, *(d) the vector forces acting on the
    system add up to zero.

8. The momentum of a system free from external forces
    is conserved (a) before and after a collision but
    not during impact between its component particles,
    (b) only if the collision is elastic (c) only if the
    collision is not elastic, *(d) always

9. Two balls moving toward each other on a frictionless horizontal surface collided and immediately came to a complete stop. This shows that the balls (a) are perfectly elastic, (b) have the same mass, (c) had equal amounts of kinetic energy before impact, *(d) had equal amounts of momentum before impact

10. Compared with its momentum before bursting, the momentum of a shell just after it burst in flight *(a) remain unchanged, (b) is changed to kinetic energy of the fragments, (c) is decreased, (d) is increased

11. Two unequal masses have the same kinetic energy. (a) The larger mass has less momentum. *(b) The larger mass has more momentum. (c) The two masses have the same momentum. (d) None of the above.

12. Two equal masses have different kinetic energies. (a) The masses have the same momentum. (b) The mass having the greater kinetic energy will have less momentum. *(c) The mass having the greater kinetic energy will have more momentum. (d) None of the above.

13. The impulse applied to an object is equal to the change in its (a) kinetic energy, (b) acceleration, *(c) momentum, (d) velocity.

14. Impulse is equal to (a) force times distance, *(b) change in momentum, (c) force divided by time, (d) mass times acceleration.

15. The units of impulse are *(a) N-s,   (b) N-m, (c) kg-s, (d) N/s

16. The units of impulse are (a) N/s, (b) m/s², (c) kg-s, *(d) kg-m/s.

17. Impulse does not depend on (a) force, (b) contact time, *(c) temperature, (d) velocity.

18. A steel ball is dropped on to a flat steel plate and the ball bounces vertically upward after collision. In this case, (a) the collision is perfectly inelastic, (b) the momentum of the ball is conserved, *(c) an impulse changes the momentum of the ball, (d) there is no external force on the ball.

19. In order to reduce the "sting" in catching a hard ball, one usually (a) increases the change in momentum, (b) increases the contact force, (c) increases the impulse, *(d) increases the contact time.

20. Padded dashboards in automobiles reduce injury by (a) increasing friction, *(b) increasing contact time, (c) decreasing friction, (d) stopping the passenger more quickly.

21. The braking action of a large jet plane after landing is chiefly due to (a) tire friction, (b) mechanical brakes, *(c) reverse thrust, (d) air resistance on wing foils.

22. When a moving ball has a head-on collision with a stationary ball of equal mass, *(a) there is a complete transfer of momentum, (b) both balls are at in motion in the same direction after collision, *(c) both balls are in motion in opposite directions after collision, (d) both balls are at rest after collision.

23. When a moving ball of small mass collides with a large (massive) stationary ball in a head-on elastic collision, (a) the large ball remains stationary, (b) the small ball is deformed, *(c) the large ball must move to conserve momentum, (d) the small ball rebounds with more momentum than it had originally.

24. When a moving massive object has a head-on inelastic collision with a small stationary object of little mass, (a) the momentum is not conserved, (b) the small mass has the same momentum as the large mass after collision, *(c) the objects may be stuck together after collision, (d) the momentum of the large mass is unchanged.

25. In a collision, *(a) momentum is exchanged or transferred, (b) contact forces are always involved, (c) momentum is not conserved, (d) the impulse is zero.

26. For an elastic collision of two objects of equal mass, (a) the total momentum is not conserved, (b) the objects are permanently deformed, (c) the impulse increases, *(d) there is a complete exchange or transfer of momentum.

27. When a moving ball of small mass strikes a
stationary ball with a large mass in an elastic
collision, (a) the total momentum decreases, *(b)
the small ball's momentum decreases and is reversed
in direction, (c) both balls move off in the same
direction, (d) the massive ball remains stationary.

28. When objects stick together after collision, (a) the
momentum is not conserved, (b) the momentum is zero,
*(c) the collision is completely inelastic, (d) the
collision is elastic.

Completion

1. <u>Momentum</u> is the product of mass and velocity.

2. The momentum of a system is not changed by <u>internal</u>
forces.

3. Momentum has magnitude and <u>direction</u>.

4. We refer to what Newton called a "quantity of
motion" as <u>momentum</u>.

5. A collection of objects defined by real or imaginary
boundaries is called a <u>system</u>.

6. An <u>external</u> force is required to change the momentum
of a body or system.

7. The momentum of an object points in the same
direction as its <u>velocity</u> vector.

8. If an external force acts on a body, its momentum
<u>changes</u>.

9. Momentum is conserved in an <u>inelastic</u> collision.

10. A change in the momentum of a mass generally
involves a change in <u>velocity</u>.

11. An internal force <u>cannot</u> change the total momentum
of a system.

12. The total momentum before a collision equals the
total momentumn after collision if there are no
<u>external</u> forces.

13. The force involved with impulse usually <u>varies</u> with
time.

14. When momentum is <u>conserved,</u> the impulse is zero.

15. The conservation of momentum is essentially expressed in Newton's <u>first</u> law of motion.

16. Momentum is conserved in a rocket system by the exhaust gases moving <u>oppositely</u> to the rocket.

17. Impulse is the product of force and <u>time</u>.

18. Impulse causes a change in <u>momentum</u>.

19. In symbol form, impulse is equal to <u>F t or mv</u>.

20. Impulse has the same units as <u>momentum</u>.

21. The "follow through" of a golf swing or base ball bat swing increases the distance the hit ball travels by <u>increasing</u> the impulse.

22. In sports, a "follow through" swing increases <u>contact time</u> giving a larger impulse.

23. Following through when hitting a golf ball sends the ball a greater distance because of increased <u>impulse or momentum.</u>

24. An interaction in which momentum is exchanged is called a <u>collision</u>.

25. Momentum is conserved in an <u>inelastic</u> collision.

26. Objects rebound without deformation from a(n) <u>elastic</u> collision.

27. A <u>collision</u> is any interaction in which momentum is exchanged or transferred.

28. Automobile air bags will reduce injuries by <u>increasing</u> collision contact time.

29. Heat may be generated in an <u>inelastic</u> collision.

30. According to the principle of conservation of momentum, if there is a net <u>external</u> force acting on a system, the momentum of a system is not conserved.

31. If two railroad cars couple together on colliding, then the collision is completely <u>inelastic</u>.

32. In a glancing collision, the total momentum is <u>conserved</u> in the component directions.

33. In an <u>elastic</u> collision, both momentum and kinetic energy is conserved.

34. In the absence of an unbalanced force, the momentum of a body is <u>conserved</u>.

35. In the absence of an unbalanced external force, the <u>total</u> momentum of a system is conserved.

Matching

(Choose the appropriate answer from the list on the right.)

g	1.	impulse	a. jet plane braking
k	2.	momentum	b. conservation of momentum
b	3.	zero net internal force	c. Newton*
m	4.	system	d. stick together
a	5.	reverse thrust	e. momentum transfer
e	6.	collision	f. internal force*
h	7.	elastic collision	g. F∆t
n	8.	inelastic collision	h. no deformation
d	9.	completely inelastic collision	i. multistage rocket
i	10.	in-flight mass reduction	j. whiplash*
			k. mv
			l. follow-through*
			m. collection of masses
			n. energy lost
			o. components*

*Answers not used.

Chapter 3

Matching

(Choose the appropriate answer from the list on the right.)

k 1. collision

m 2. internal force

l 3. conservation of momentum

e 4. elastic collision

c 5. momentum

n 6. inelastic collision

o 7. total momentum

a 8. external force

f 9. system

i 10. impulse

a. outside force

b. force/area*

c. mass x velocity

d. constant mass*

e. no energy lost

f. collection of objects

g. inertia*

h. velocity*

i. $F\Delta t$

j. $\frac{1}{2}mv^2$

k. exchange of momentum

l. no change in momentum

m. force from within

n. energy lost

o. vector sum of momenta

*Answers not used.

Chapter 4   Projectile, Circular, and Planetary Motion

Answers to Questions

1. (a) and (b). The same in both cases, since independent of mass.

2. Coin would rise and fall, outside observer would see a parabolic path.

3. $0°$

4. Object has same horizontal velocity as airplane when dropped. The air resistance on the plane and object would be comparable.

5. (a) parabola, (b) an elongated path with accelerations in both the x and y directions.

6. (a) shorter height, (b) and (c) different arcs due to air resistances in vertical and horizontal directions.

7. The plane was piloted (dived) under the arc of the bullet path.

8. Yes, if the vertical component of velocity of the projection at an angle is the same as that of the vertical projection.

9. It would fall downward and not reach home plate.

10. Yes, $55°$, which would have a greater maximum height.

11. Discus thrower, javelin thrower, in artillery firing, etc.

12. $a_c = v^2/r = (m/s)^2/m = m/s^2$

13. Necessary friction in washer. In dryer, the force due to gravity is greater than the centripetal force, which depends on the rotational speed.

14. $F_1 = mv^2/r$, $F_2 = m(3v)^2/r = 9F_1$, due to friction.

15. A component of the nonvertical reaction force supplies centripetal force.

16. Technically no, it is not an inertial system since the Earth is rotating.

17. They would slide down the wall.

Chapter 4

18. Due to an outward "centrifugal" force.

19. The heavier milk.

20. As a projectile travels south, the Earth would rotate and the projectile would land to the west (or right) of a north-south line (meridian). In the southern hemisphere, by the same reasoning a north-bound projectile would be deflected to the left.

21. Slow down the colony rotation so the centripetal force would be less.

22. Tacks close together, a circle. Tacks far apart; a straight line.

23. (a) The slowest speed is in July when the Earth is the greatest distance from the Sun.
(b) The fastest speed is in January when the Earth is closest to the Sun.

24. The speed would be constant.

25. The greater the distance of the planet from the Sun the greater the period. Kepler's third law:

$T^2 = kR^3$, when R increases T increases.

26. $T^2 = kR^3$, or $(1 \text{ year})^2 = k(1 \text{ AU})^3$, and

$k = (1 \text{ year})^2/\text{AU}^3$, so $T^2 = R^3$ with these units.

SAMPLE TEST QUESTIONS

Multiple Choice

1. A vertically projected object (a) has zero acceleration at maximum height, (b) has a constant velocity, *(c) returns to its starting point with the same initial speed, (d) has a greater acceleration than a horizontally projected object.

2. The horizontal acceleration of an object projected at an angle, (a) increases with time, *(b) is zero, (c) is the same as the vertical acceleration, (d) is greater when projected at 45°.

3. A projectile *(a) has a constant speed in the horizontal direction (b) is always projected in one dimension, (c) has no forces acting on it, (d) has no vertical acceleration.

4. For which projection is the horizontal acceleration zero? (a) vertical projection, (b) horizontal projection, (c) projection at an angle, *(d) all of the preceding.

5. The type of path an object projected at an angle (other than $90°$) follows *(a) is a parabolic arc, (b) is a circular arc, (c) depends on its initial velocity, (d) depends on the angle of projection.

6. If air resistance is a factor in a horizontal projection or a projection at an angle, the range of the projectile would be (a) greater, *(b) less, (c) the same.

7. A mass projected vertically upward with an initial velocity v_i has (neglecting air resistance).
   (a) a velocity vector directed vertically downward,
   (b) an acceleration vector in the upward direction,
   *(c) both a velocity and acceleration vector which are in opposite directions,
   (d) none of the above.

8. A rifle bullet which has been projected horizontally with an initial velocity v_i has (neglecting air resistance)
   (a) a horizontal acceleration,
   *(b) an acceleration vertically downward,
   (c) a range independent of the original height of the rifle above ground level,
   (d) all of the above.

9. A cannonball projected at a 45 degree angle with an initial velocity v_i (neglecting air resistance) reaches
   (a) maximum horizontal velocity at maximum height,
   (b) minimum acceleration at maximum height,
   (c) minimum horizontal velocity at maximum height,
   *(d) maximum acceleration at maximum height.

10. A cannonball is projected at at 45 degree angle. At maximum height of the cannonball the direction for the velocity and acceleration are (a) in the same direction, (b) in opposite directions, *(c) 90 degrees to one another, (d) 45 degrees to one another.

11. A piece of chalk is projected horizontally from the top edge of a table. At the same instant a piece of chalk is dropped from the same table height.
    (a) the projected chalk will reach the floor first,
    (b) the dropped chalk will reach the floor first,
    *(c) both will reach the floor at the same time,
    (d) none of the above.

12. A piece of chalk is projected horizontally from the top edge of a table. At the same instant a piece of chalk is dropped from the same table height.
    (a) both will have the same velocity when they reach the floor, (b) the dropped chalk will have the greater velocity, *(c) the projected chalk will have the greater velocity, (d) the dropped chalk will reach the floor first.

13. A ball is projected upward at a 45 degree angle. If we neglect air resistance, the motion of the ball consists of a uniform downward acceleration combined with (a) a constantly increasing velocity, *(b) a uniform horizontal velocity, (c) a uniform vertical velocity, (d) none of these.

14. Uniform circular motion requires, (a) a tangential force, *(b) a tangential velocity, (c) a constant velocity, (d) zero acceleration.

15. The centrifugal force (a) is the third law reaction force to centripetal force, (b) is a center-seeking force, *(c) does not exist in an inertial system, (d) is necessary for circular motion.

16. An object in uniform circular motion has a constant (a) velocity, (b) tangential acceleration, (c) momentum, *(d) speed.

17. The centripetal force acting on a car making a turn on a level road is supplied by (a) the weight of the car, *(b) friction between the tires and the surface of the road, (c) gravity, (d) all of these.

18. An object in uniform circular motion has (a) a constant velocity, *(b) a constant speed, (c) zero acceleration, (d) an acceleration radial outward.

19. A centrifuge separates particles of different (a) mass, (b) momentum, *(c) density, (d) velocity.

20. All planets of the solar system move in _____ orbits about the Sun. (a) circular, (b) parabolic, *(c) elliptical, (d) none of these.

21. In uniform circular motion, the _____ of an object is not constant. (a) speed, (b) acceleration, *(c) velocity, (d) none of these.

22. The laws of planetary motion were developed by (a) Newton, (b) Galileo, (c) Brahe, *(d) Kelper.

23. The shape of planetary orbits is given by Kepler's _____ law *(a) first, (b) second, (c) third, (d) fourth.

Completion

1. An object that has been thrown is called a projectile.

2. A vertical projectile has an initial upward velocity and a downward acceleration.

3. A projectile launched at an angle follows a parabolic path.

4. For a vertical projectile traveling upward, the velocity and acceleration vectors are in the opposite direction.

5. In a vacuum, a vertical projectile returns to its starting point with a speed equal to its initial speed.

6. The angle used to obtain the maximum range for a projectile launched at ground level is 45 degrees.

7. A horizontal projection has acceleration due to gravity.

8. In uniform circular motion, an object will have constant speed.

9. In circular motion, the change in the velocity has an instantaneous direction toward the center of the circle.

10. A centripetal acceleration is required for circular motion.

11. The centripetal acceleration of an object in uniform circular motion depends on the object's orbital speed.

12. The centripetal force required for a car to round a circular curve depends on the speed of the car and the <u>radius of curvature</u>.

13. To increase simulated gravity in a rotating space colony, the rotational speed would be <u>increased</u>.

14. A force on an object at a right angle to the radius of its circular motion is known as a <u>tangential</u> force.

15. A centripetal force is directed toward <u>the center</u> of a circular path.

16. If the centripetal force on an object in circular motion went to zero, the object would fly off in a direction <u>tangential</u> to the circle.

17. Centrifugal force is sometimes called a <u>pseudo</u> force.

18. The centripetal force for a car to round a circular curve is supplied by <u>friction</u>.

19. Kepler's law of equal areas states that an imaginary line from a planet to the Sun sweeps out equal areas in equal <u>times</u>.

20. The period of a planet's orbit depends on its <u>distance</u> from the Sun.

21. According to Kepler's law of <u>elliptical orbits</u>, all planets move in elliptical paths about the Sun.

22. A planet has a greater orbital speed when it is <u>closer</u> to the Sun.

23. The orbits of the planet's have the shapes of a/an <u>ellipse</u>.

Matching

(Choose the appropriate answer from the list on the right.)

_d_ 1.	projectile	a. pseudo force
_j_ 2.	horizontal projection	b. equal areas
_g_ 3.	vertical projection	c. directly proportional to mass
_f_ 4.	range	d. thrown object
_i_ 5.	maximum range	e. elliptical orbits
_l_ 6.	projection at an angle	f. horizontal distance
_k_ 7.	centripetal acceleration	g. $v_x = 0$
_c_ 8.	centripetal force	h. orbital periods
_a_ 9.	centrifugal force	i. 45 degrees
_m_ 10.	uniform circular motion	j. zero horizontal acceleration
_o_ 11.	tangential velocity	k. proportional to v^2
_n_ 12.	centrifuge	l. original vertical velocity not zero
_e_ 13.	Kepler's first law	m. constant speed
_b_ 14.	Kepler's second law	n. separation by densities
_h_ 15.	Kepler's third law	o. right angles to radius

Answers to Questions

1. They are the most basic or "fundamental" forces known.

2. It is moving tangentially at the same time giving rise to circular motion.

3. The same, mg. by Newton's 3rd law, in the opposite direction.

4. With $r_2 = 2r_1$, $F_2/F_1 = (r_1/r_2)^2 = (1/2)^2 = 1/4$, and $F_2 = F_1/4$.

5. Pi, π

6. Technically, there is an accelerating force on approach and a decelerating force after passing, but the masses are too small for forces to have appreciable effects.

7. The Earth's attraction is much greater $(1/r^2)$.

8. It would fly apart.

9. Stone would fall back and forth (oscillate) through center of Earth. If the hole were through the center of the Earth, it would come out in the Indian Ocean -- not China.

10. Two high and low tides are observed daily because of the gravitational attraction between the moon and the Earth, and because the Earth makes one rotation daily. The gravitational force of attraction between two masses is inversely proportional to the distance between the two masses. Thus, the near side of the Earth to the moon has the greater force of attraction and the far side less attraction. When the sum of all forces are taken in reference to the center of the Earth, the resultant forces acting produce one bulge of the Earth closest to the moon and another bulge opposite the moon with minimum forces at 90 degrees with the two bulges. The rotating Earth positions the bulges so that two high and two low tides occur daily.

11. Because of tidal friction with the Earth and land forms.

12. Spring tides are higher tides than normal. They
    take place at the time of new and full phase of the
    moon. At these times the Sun, moon, and Earth are
    positioned in the same plane (viewed from above).
    Neap tides are high tides but are not as high as
    normal. They take place at the time of first
    quarter and last quarter phase of the moon. At
    these times the moon is 90 degrees east and 90
    degrees west of the Sun's position.

13. Full moon, when the Sun is on the opposite side of
    the Earth.

14. Tidal range is the difference in height between high
    and low tides. Varies from 1-2 m in open ocean to
    16 m in Bay of Fundy. Also affected by spring and
    neap tides.

15. Very small tides or rise of solid Earth.

16. Indirectly, no variations in the orbits of Venus and
    Mars. Directly, send a picture-taking probe so as
    to view region on other side of the Sun.

17. Planet would be very distant and very faint.
    Difficult to pin-point or see. Recall planets are
    visually observed by reflected sunlight.

18. At the time of grand alignment the Sun and the
    planets are nearly in the same plane which allows
    the addition of the vector forces of gravitational
    attraction to have greater values.

19. $g' \propto 1/(4R_e^2) = (1/16)R_e^2 = g/16$ (Note: distance from
    Earth's center is $4R_e$.)

20. Mass and radius smaller.

21. g and weight on moon is 1/6 that on Earth, and 300
    lb/6 = 50 lb.

22. Yes, different masses and radii.

23. No. The mass of a planet.

24. Much larger or greater radius.
    $g = GM_e/R_e^2 = G(94M_e)/R_s^2$, and $R_s = (94)^{\frac{1}{2}}R_e =$
    $(9.7)R_e$

25. Field lines concentrated in region between masses,
    see text.

26. (a)   Field lines with stronger field near greater mass. (b)   Greater field "depression" near Earth.

27. The region outside the sphere with the radius of the original star. (Same mass within both.)

28. Smaller radius.

29. Infinite distance.

30. No, gravity is acting because you are falling.

31. The weight of an object is defined as the attractive force between the Earth and the object. The gravitational force is present, therefore the water has weight. During free fall the water and cup have weight or gravitational attraction and fall at the same rate.

32. The scale would indicate a larger reading due to the upward acceleration of the elevator.

33. (a) Downward force (acceleration) or heavier. (b) Upward force (acceleration) or lighter. Similar on rocket.

34. Wild. Ball and players "floating" around.

35. (a) Enough tangential speed to keep in orbit.   (b) It would fall out.

36. No. Gravity less, minimum velocity less.

37. No. Less because gravity less. No atmosphere, no air resistance.

38. No. Different masses and radii.

39. Less gravity, atmospheric gases were eventually lost.

SAMPLE TEST QUESTIONS

Multiple Choice

1. The fundamental force associated with mass interaction is (a) electromagnetic, *(b) gravity, (c) strong nuclear, (d) weak nuclear.

2. The weakest fundamental force is the (a) electromagnetic force, *(b) gravitational force, (c) weak nuclear force, (d) strong nuclear force.

3. The constant G is (a) the acceleration due to gravity, (b) smaller on the moon than on Earth, (c) the force of gravity, *(d) a universal constant.

4. The force of gravity (a) does not depend on the masses, (b) acts only between large masses, *(c) is inversely proportional to the square of the separation distance, (d) is always repulsive.

5. The gravitational constant G is (a) the same as g, (b) is a very large number, (c) changes with time, *(d) is believed to be universally constant.

6. Lunar gravity gives rise to (a) coments, *(b) tides, (c) apparent weightlessness, (d) microgravity.

7. A gravitational field *(a) is the force per unit mass at points in space, (b) does not depend on the size of a mass, (c) causes apparent weightlessness, (d) is always directed away from a mass.

8. The units for the acceleration of gravity are (a) km/s, *(b) km/s^2, (c) km/s/km, (d) $km/s^2/km$

9. The units for the universal gravitational constant are (a) km/s, (b) $n-kg^2/m^2$, *(c) $n-m^2/kg^2$, (d) $n-m^2/kg$

10. The distance between the centers of two spherical masses on the classroom lecture desk is one meter. The gravitational force of attraction between them is F. The masses are moved farther apart so that the distance between their centers is four meters. The force of attraction between the two masses has been reduced to (a) F/4, (b) F/8, *(c) F/16, (d) F/32.

11. Consider the moon's orbit around the Earth to be circular and suppose the moon's mass to be doubled. Compared with the present force of attraction between the moon and the Earth, the new attractive force between them would be *(a) doubled, (b) one-half as much, (c) four times as much, (d) one-fourth as much.

12. Higher high tides and lower low tides caused by the added pull of the Sun at the times of new and full moon are called (a) neap tides, (b) summer tides, *(c) spring tides, (d) Sun tides

13. A couple living at the beach would notice on the average how many spring tides a month? (a) none, (b) one, *(c) two, (d) four.

14. The universal gravitational constant (G) is _____
to the acceleration due to gravity (g). (a) inversely proportional, *(b) directly proportional,
(c) greater than, (d) the same as.

15. At an altitude equal to the Earth's radius, a person
would weigh what percentage of his or her weight on
Earth?   (a) 200%,   (b) 100%, (c) 50%, *(d) 25%.

16. Weightlessness experienced in an orbiting Earth
satellite (a) is due to lunar gravity, (b) results
from zero gravity, *(c) arises as a result of being
in orbit, (d) is due to a gravitational field.

17. To put a satellite in a circular orbit about the
Earth requires a minimum tangential velocity of (a)
2 km/s, (b) 4 km/s, (c) 6 km/s, *(d) 8 km/s.

18. If a projectile were given an initial vertical
launch speed of 11 km/s or greater, it would *(a)
escape from Earth, (b) go into circular orbit, (c)
go into an elliptical orbit, (d) fall back to Earth.

19. An astronaut in a spaceship in a circular orbit
about the Earth (a) is weightless, *(b) has less
than one g of force acting on him, (c) has a
centripetal force supplied by the spaceship, (d) has
an orbital speed of 11 km/s.

20. An Earth satellite in circular orbit at an altitude
of several hundred kilometers would have an orbital
speed of *(a) less than 8 km/s, (b) about 8 km/s,
(c) 9 km/s, (d) 11 km/s

21. An astronaut in a spaceship in a circular orbit
about the Earth wishes to put the ship into an
elliptical orbit. To do so, he could *(a) apply a
reverse or forward thrust with the rocket engines,
(b) go on space walk, (c) make the ship lighter by
ejecting equipment, (d) use the combined effect of
lunar and solar gravity.

22. About what year will Halley's comet make its first
appearance in the 21st century? (a) 2001, (b) 2048,
*(c) 2062, (d) 2085.

Completion

1. Newton formulated the law of gravitation.

2. The constant G of the law of gravitation is called the <u>universal gravitational constant</u>.

3. A <u>gravitational field</u> is used to represent the gravitational effects around a mass.

4. Einstein perceived a gravitational field as a warping of four-dimensional <u>space</u> and <u>time</u>.

5. The law of gravitation states that the attractive force is <u>directly</u> proportional to the magnitude of the masses.

6. The law of gravitation states that the attractive force is <u>inversely</u> proportional to the square of the distance between the masses.

7. Gravity is always a/an <u>attractive</u> force.

8. The units for the acceleration of gravity (g) in the MKS system of units is <u>km/s²</u>.

9. The units for the universal gravitational constant (G) in the mks system of units is <u>N-m²/kg²</u>.

10. The quantity (g) is the <u>acceleration</u> of gravity.

11. <u>Gravity</u> is the weakest fundamental force of nature.

12. The force that considers the masses of two objects and their separation distance is the <u>gravitational force</u>.

13. Every particle is <u>attracted</u> to every other particle.

14. The magnitude of the acceleration due to gravity near the Earth's surface is <u>9.8 m/s²</u>.

15. Apparent weightlessness is sometimes incorrectly called a state of <u>zero</u> gravity.

16. Apparent weightlessness and other effects on Earth-orbiting spacecraft is collectively called <u>microgravity</u>.

17. To place a satellite in elliptical orbit about the Earth requires a tangential velocity greater than <u>8 km/s</u>

18. Instead of giving a spacecraft an initial escape velocity to leave the Earth, <u>multistage</u> rockets are used.

19. Two "g's" of force acting on a person would be <u>two</u> times the person's weight.

20. The acceleration due to gravity near the Earth varies slightly with <u>altitude</u>.

21. Tides are not equally high due to the <u>inclination</u> of the moon's orbit.

22. Neap tides occur during the <u>1st and 3rd</u> quarter moons.

23. The gravitational attraction of the Sun gives rise to <u>spring</u> tides at the times of new and full moons.

24. <u>Centripetal</u> force causes the moon to "fall" toward the Earth.

25. Spring tides result when the <u>Sun</u> is on the same side or the opposite side of the Earth from the moon.

26. Ocean tides are primarily due to the <u>moon</u>.

27. <u>Neap</u> tides are a condition of lower high tides.

28. The "tidal" friction between the oceans and the oceans' floors slows down the Earth's rotation about 1/1000 of a second per <u>century</u>.

29. To put a satellite into orbit, gravity must first be overcome and then the satellite given a sufficient <u>tangential</u> velocity.

30. Astronomers are searching for a possible tenth planet which they call <u>Planet X</u> .

Matching

(Choose the appropriate answer from the list on the right.)

<u>e</u> 1. Newton's law of gravitation

<u>g</u> 2. Universal gravitation constant

<u>m</u> 3. Periodic rising and lowering of the oceans

<u>o</u> 4. Spring tides

<u>i</u> 5. Neap tides

<u>b</u> 6. Circular orbit

<u>j</u> 7. Gravitational field

<u>c</u> 8. Weightlessness

<u>h</u> 9. 2 g's

<u>n</u> 10. Escape velocity

<u>f</u> 11. Elliptical orbit

<u>k</u> 12. Acceleration of gravity

<u>a</u> 13. Fundamental force

<u>l</u> 14. Microgravity

<u>d</u> 15. Orbital decay

a. gravitation

b. about 8 km/s

c. absence of a reaction force

d. drag from the atmosphere

e. $F = Gm_1 m_2 / r^2$

f. greater than 8 km/s

g. $N\text{-}m^2/kg^2$

h. twice a body's weight

i. occur when moon is at first or last quarter phase

j. space-time warp

k. g

l. infinitesimal forces

m. tides

n. 11 km/s

o. occur when moon is at new or full phase

Chapter 6 Rotational Motion

Answers to Questions

1. No. In a uniform field location of the c.g. is the
    same as the c.m. but, even in a nonuniform field,
    both exist.

2. Between the arms of the boomerang. Location can be
    determined by suspension.

3. Yes, but not in normal positions, otherwise the
    person would topple over.

4. (a) In bending the body, the c.g. may actually be
    below the bar. (b) The energy required is less
    since the c.g. is not raised as high.

5. No. Two points are sufficient, but it may be
    convenient to use more points.

6. Directly below the main suspension string.

7. No. Cannot do both pure motions at once by
    definition.

8. The chance that the ball will have pure
    translational motion is extremely small. It is
    possible, but highly improbable.

9. The wheels of an automobile undergo a rolling
    motion. That is, the wheels rotate about an axis
    that moves translationally while opposite points on
    the wheel has opposite tangential velocities.

10. A combination of rotational and translational
    motions.

11. (a) 2π rad, (b) π rad, (c) $\pi/2$ rad

12. (b) 2π rad/1 s = 2π rad/s, (b) 2π rad/1 min =
    2π rad/min, (c) 2π rad/1 h = 2π rad/h

13. Angular speed is same anywhere, e.g., 33-1/3 rpm.
    Tangential speed, $v = rw$ decreases as r decreases.

14. (a) they both have the same angular speed, (b) The
    child near the outer edge has the greater tangential
    speed.

15. No. Outer band smaller, since greater recording
    distance per rotation.

16. Fair ruling, since larger tires would give greater distance per revolution and greater speed. If odometer were based on revolutions, then size of tires would make no difference.

17. Yes, in as much as $w = \alpha t$ and the direction of the angular acceleration affects the change in α .

18. Wheel has large moment of inertia, and small impulses of energy are not noticeable.

19. (a) Very small. (b) Large, more mass farther from axis. (c) Small, since mass near axis of rotation.

20. Torque and rotation about front end.

21. Yes, equal and opposite forces at different distances from axis of rotation.

22. Maximum torque when pedals and sprocket arm (lever arm) horizontal. Minimum torque when sprocket arm vertical. Minimum torque is zero when force is through axis of rotation, but pedals swivel to prevent this, particularly from the foot and pedal at top of sprocket revolution.

23. The screwdriver with the larger handle. Being a wheel and axle, the mechanical advantage is given by diameter of handle/diameter of shaft.

24. (a) The larger wheel gives greater torque due to the greater radius. (b) No. The response would be excessive for the massive truck.

25. The vector summation of the torques on each side of the balance point is zero, so there is no rotational motion.

26. When balanced, torques are equal and no rotational motion.

27. $F_2 = 4F_1$, $F_1 r_1 = F_2 r_2$, and $r_1 = (F_2/F_1)r_2 = 4r_2$.

    With $r = r_1 + r_2 = 100$ cm, $r_1 = 80$ cm and $r_2 = 20$ cm.

28. (a) Inside the Earth. (b) Inside the Sun. (c) Inside the Sun.

29. Loss of potential energy by center of gravity.

30. Yes, in the sense that forces are equal and opposite, but the object acted on by the reaction force may not have an axis of rotation.

31. Greater base and greater stability.

32. Higher center of gravity, greater moment of inertia, and slower rotation. Allows more time for response.

33. Lower center of gravity and more stability.

34. Center of gravity is always such that there is a restoring (righting) torque.

35. The twelfth brick (12 bricks total counting base brick) would put the center of gravity at the edge of the base of support. Being in unstable equilibrium, the stack would probably fall. For certain with the addition of another brick.

36. It depends on how far the c.g. is inside of the base of support in each case and its height.

37. (a) Laying down. (b) Forward extension of center of gravity and helping torque.

38. More unstable, raises, c.g.

39. Greater base of support -- in lateral direction (side ways only).

40. Yes. The higher the load, the less stable and if not evenly (horizontally) distributed, the greater torque going around curves.

41. (a) When c.g. falls outside of base of support. (b) Put weights on opposite side of c.g.

42. Stable -- 6 different sides. Unstable -- 12 edges and 8 points (corners)

43. The handles of the fork and spoon move the c.g. back so it is above the chase of support.

44. No. The angular momentum of a particle is a function of the radial distance from the axis of rotation. The greater the radius the greater the angular momentum.

45. The high diver decreases the radial distances of the masses of his or her body by pulling in to a more compact shape thus increasing the angular velocity.

46. The turntable would rotate oppositely to conserve angular momentum. Rate depends on moment of inertia. Spiral path would increase speed because of decrease in radial distance.

47. Stablizing torques applied to wings and tail.

48. The angular momentum of the rotating wheels provides the stability.

SAMPLE TEST QUESTIONS

Multiple Choice

1. A "particle" does not have (a) mass, (b) an exact position, *(c) physical dimensions, (d) motion.

2. The center of gravity is (a) always at the center of a body, *(b) sometimes outside a body, (c) always where the body is most dense, (d) not described by any of the preceding.

3. The axle of a wheel of an automobile moving on a level surface travels in a straight line. The wheel then has (a) pure translational motion, (b) pure rotational motion, *(c) a combination of translational and rotational motions, (d) neither translational or rotation motions.

4. All of the particles of a second hand of a clock have (a) an angular speed of $\pi/30$ rad/s, *(b) the same angular speed, (c) the same tangential velocity, (d) different angular accelerations.

5. Two bar bells of the same length have different weights at their ends, but equal end weights on the same bar. The bar bells have (a) different locations for their centers of gravity, (b) the same rotational inertia, *(c) different moments of inertia, (d) the same angular acceleration for the same torque.

6. A net torque *(a) produces rotational motion in every instance, (b) always acts about an axis through the center of gravity, (c) always produces clockwise rotations, (d) can occur in the absence of a force.

7. Which of the following sets is <u>not</u> translational and rotational analogs. (a) x and θ, (b) m and I, (c) ma and rF, *(d) mv and Iw.

8. Considering a textbook to have a rounded binding end or surface (which it usually does), how many different positions of stable equilibrium does it have? (a) 4, *(b) 5, (c) 6, (d) 7.

9. If the angular speed of a rotating object is doubled, its rotational energy is (a) doubled, (b) tripled, *(c) quadrupled, (d) not changed.

10. If the angular momentum of a system is conserved, (a) its angular velocity is always constant, (b) its moment of inertia is always constant, (c) both the angular velocity and the moment of inertia are always constant, *(d) the product of the angular velocity and moment of inertia is constant.

11. The average location of the weight distribution of a body is called (a) rotational inertia, (b) center of mass, *(c) center of gravity, (d) lever arm.

12. For a body in pure translational motion, (a) the location of the center of mass is shifted to a new position in the body, *(b) the angular momentum is conserved, (c) particles of the body rotate in circles, (d) there is no moment of inertia.

13. A spiraling football is an example of (a) pure translational motion, (b) pure rotational motion, (c) a nonrigid body, *(d) the general motion of a rigid body.

14. The number of radians in one-half of a complete rotation is (a) $\pi/4$, (b) $\pi/2$, *(c) π, (d) 2π.

15. In the absence of an unbalanced torque, a rotating object moves with *(a) a constant angular velocity, (b) an increasing angular acceleration, (c) in pure translational motion, (d) a force and a lever arm.

16. Rotational inertia (a) does not depend on the object's mass distribution, *(b) is greater if the mass distribution is farther from the axis of rotation, (c) depends on the torque acting on a body, (d) determines stability.

17. The unit of torque is (a) joule, (b) lb, (c) N, *(d) m-N.

18. An object with its center of gravity outside its base of support (a) does not experience a torque, (b) is in stable equilibrium, *(c) is in unstable equilibrium, (d) has a constant angular velocity.

19. Rotational kinetic energy *(a) depends on the mass distribution of a body, (b) is equal to Iw, (c) requires the conservation of angular momentum, (d) is the same as translational kinetic energy.

20. An ice skater goes into a rapid spin on the end of one skate blade because of (a) constant angular velocity, *(b) the moment of inertia is changed, (c) the angular momentum is not conserved, (d) the change of direction of the angular momentum.

21. The point where all the mass of a body can be considered to be concentrated is known as the center of (a) gravity, *(b) mass, (c) weight, (d) torque

22. The time rate of change of angular velocity is known as angular (a) momentum, (b) rotational velocity, *(c) acceleration, (d) inertia.

23. The perpendicular distance from the axis of rotation to a line along which a force acts is known as the (a) torque, (b) moment of inertia, (c) force distance, *(d) lever arm.

24. Newton's second law of rotational motion states that torque equals (a) moment of inertia x angular velocity, (b) angular acceleration, *(c) moment of inertia x angular acceleration, (d) mass x acceleration.

25. The condition of a body when its center of gravity is vertically above an edge or point such that a slight displacement will cause the body to topple is known as a condition of _____ equilibrium. (a) stable, *(b) unstable, (c) neutral, (d) rigid

26. The units for angular speed in the MKS system of units are (a) m/s, (b) km/s, *(c) radians/s, (d) none of these.

27. The units for angular acceleration in the mks system of units are (a) radians/s, (b) m/s^2, (c) degrees/s, *(d) radians/s^2.

28. The units for angular momentum in the MKS system of units are *(a) kg-m^2/s, (b) kg-rad/s, (c) kg-m/s, (d) kg-m^2/s^2.

29. Snow tires have a larger diameter than regular tires. With snow tires on your car a speedometer reading of 55 mi/h would be (a) an accurate reading of the car's speed, (b) a value greater than the speed of the car, *(c) a value less than the speed of the car, (d) none of these.

30. One radian is a unit of angular measure equal to (a) 360/2, (b) 360/π, (c) 67.3 degrees, *(d) arc length equal to the radius/radius of circle.

Completion

1. In pure rotational motion, the particles of a body move in circles about a line called the <u>axis of rotation</u>.

2. Angular velocity is generally measured in units of <u>rad/s</u> or <u>rev/min</u>.

3. The rotational inertia of a body is expressed in terms of its <u>moment of inertia</u> about an axis of rotation.

4. The rotational analog of force is <u>torque</u>.

5. A wheel slows uniformly from a rotational speed of 10 rad/s to rest in 5 seconds. The magnitude of the angular acceleration is <u>2 rad/s</u>.

6. A meter stick is balanced at its center on a support. A 100 gram mass is placed on the 20 cm mark. Another 100 gram mass would be placed on the <u>80</u> cm mark to produce rotational equilibrium or zero torque.

7. A ball or cylinder which rolls down an inclined plane has <u>rotational</u> and <u>translational</u> kinetic energies.

8. If there is no net torque acting on a body, its <u>angular momentum</u> is conserved.

9. The unit of torque in the SI system is <u>m-N</u>.

10. When a force acts through the axis of rotation of a body, the torque is zero because the lever is equal to <u>zero</u>.

11. The average location of all the mass particles that make up an object is called the <u>center of mass</u>.

12. The center of gravity of a flat, irregularly shaped object may be located by means of <u>suspension</u>.

13. The center of mass of an object moves as though it were a <u>particle</u>.

14. The unit of angular speed is <u>rad/s</u>.

15. Rotational inertia depends on the <u>mass distribution</u> about the axis of rotation.

16. Circus tightwire walkers often carry a long pole in order to increase the <u>rotational inertia</u>.

17. Angular acceleration requires <u>a net torque</u>.

18. If a body in stable equilibrium is slightly displaced, a <u>(restoring) torque</u> tends to bring it back to its equilibrium position.

19. A solid cylinder will roll <u>faster</u> than a hollow cylinder down an inclined plane.

20. Angular momentum is conserved in the absences of <u>a net torque</u>.

21. The center of mass for a uniform doughnut is located at the <u>center</u> of the doughnut.

22. A photograph which turns at a rate of 33-1/3 rpm completes 100 revolutions in <u>3</u> minutes.

23. A frictional force is applied to a bicycle wheel which is rotating. The angular acceleration that results would be <u>negative</u>.

24. A mechanic tries to loosen a bolt by placing an extension on the handle of the wrench. In so doing, he is increasing the torque by increasing the <u>lever arm</u>.

25. If a diver's rotational inertia is decreased by one-half when she goes into a "tuck" position, her angular velocity is increased <u>twice</u> the initial amount.

26. Ten revolutions is equivalent to <u>20</u> π radians.

27. The tangential speed of a particle on a rotating wheel is greatest on the <u>rim</u> of the wheel.

28. A solid cylinder has twice the mass of a hollow cylinder with the same outer radius. Both cylinders start rolling from rest from the top of an inclined plane at the same time. The <u>solid</u> cylinder will reach the bottom of the plane first.

29. Angular acceleration is a <u>vector</u> quantity.

30. In a uniform gravitational field, the center of mass and the center of gravity are at the <u>same location</u>.

31. In pure translation motion, the particles of a body move with the same <u>instantaneous velocity</u>.

32. Angular distance is expressed in <u>radians</u>.

33. Rotational inertia is expressed in terms of the <u>moment</u> of inertia.

34. The rotational analog of force is <u>torque</u>.

35. An object in stable equilibrium has its center of gravity <u>over</u> its base of support.

36. Large helicopters have two rotors that rotate in the <u>opposite</u> direction to conserve angular momentum.

37. Large helicopters have two rotors that rotate in the opposite direction to conserve <u>angular momentum</u>.

38. Newton's first law for rotational motion states that a rigid body remains at rest or in motion with constant angular velocity unless acted upon by an unbalanced <u>torque</u>.

39. In the absence of an unbalanced <u>torque</u>, the total angular momentum of a system is <u>conserved</u>.

Matching

(Choose the appropriate answer from the list on the right.)

c	1. rotational motion	a. average location of the total mass of a body
d	2. angular speed	b. greater lever arm
g	3. angular velocity	c. motion about an axis of rotation
e	4. angular acceleration	d. time rate of change of angular distance
m	5. angular momentum	e. time rate of change of angular velocity
k	6. rotational kinetic energy	f. collection of fixed particles
n	7. tangential speed	g. radians/s + direction
a	8. center of mass	h. conditions when net torque is zero
p	9. rotational inertia	i. depends on location of center of gravity
b	10. increased torque	j. $\frac{1}{2}mv^2 + \frac{1}{2}Iw^2$
j	11. total kinetic energy of a rolling body	k. $\frac{1}{2}Iw^2$
l	12. torque	l. $I\alpha$
h	13. rotational equilibrium	m. Iw
f	14. rigid body	n. rw
o	15. angle passed through	o. $\theta = wt$
i	16. stability	p. I

Chapter 7 Atoms, Molecules, and Matter

Answers to Questions

1. The impulses of the molecular collisions are small and random, and so cancel.

2. Pictures of atoms.

3. U.S. population approximately 2.5×10^8 persons.

    World population about 5.0×10^9 persons. Molecules in 1/2 liter of air on the order of 10^{22}.

4. (a) 6 L (10^{22} mcls/1/2L) = 12×10^{22} mcls.
    (b) The 78% of the N_2 molecules are unchanged, but CO_2 is expelled in place of O_2 (oxygen and carbon dioxide exchange takes place in the lungs, O_2 in the air, CO_2 in blood).

5. Molecules in one liter on the order of 10^{25}.
    National debt on the order of 10^{12} (trillion) dollars.

6. (a) similar, (b) similar, (c) different, (d) different, (e) different.

7. Yes, if it is a neutral atom and not an ion.

8. Each have two "extra" (valence) electrons outside a filled shell.

9. Na has one valence electron and F lacks one electron for a filled shell.

10. (a) silver, (b) gold, (c) mercury, (d) iron

11. (a) $_{29}Cu$, (b) $_{90}Th$, (c) $_{102}No$

12. A metal with similar properties since Pd is in same group.

13. A thorium atom.

14. No, the definition refers to chemical properties and character. More than a few sugar molecules are needed to reach the taste threshold. Think of what it would be like if we "tasted" every molecule of things we eat.

15. $C_{12}H_{22}O_{11}$ --> 12 C + 11 H_2O, so with 12 sugar molecules, 144 C and 132 H_2O.

16. The solution is a mixture of water and salt that can be separated by physical evaporation so there is no chemical change involved, only a physical one.

17. Temperature and pressure.

18. Yes. solid to liquid ---- melting
           solid to gas ------ sublimation
           liquid to gas ------ evaporation
           liquid to solid ---- freezing
           gas to liquid ------ condensation
           gas to solid ------- sublimation (deposition)

19. Molecules are held together by intermolecular forces (electrical) bonds. The addition of heat (energy) to a substance causes the molecules to increase their motion about their equilibrium position. When sufficient energy is added the bonds are broken and a phase change takes place.

20. (a) solid, (b) solid and liquid

21. They are quickly annihilated.

22. Similar to regular atoms with protons replaced by antiprotons and electrons by positrons.

23. Handling problem might be solved, but the energy input to create the antimatter in the engine would essentially double the engine (work) input and make it very inefficient. The energy used to create the antimatter could be used directly in an engine.

24. It would be annihilated in the atmosphere before it could land.

SAMPLE TEST QUESTIONS

<u>Multiple Choice</u>

1. Who supported the continuous theory of matter? (a) Democritus, (b) Gassendi, (c) Newton, *(d) Aristotle.

2. Brownian motion was explained by (a) Newton, (b) Brown, (c) Dalton, *(d) Einstein.

3. The diameters or widths of atoms are on the order of (a) 10^{-10} m, (b) 10^{-8} cm, (c) 1 $\overset{\text{o}}{\text{A}}$, *(d) all of the preceding.

4. The electrons in an atom (a) contribute significantly to the mass of the atom, *(b) are in specific orbits or shells around the nucleus, (c) determine the atomic species, (d) are repelled by the nuclear protons.

5. The atomic species or kind of atom is determined by the (a) mass number, *(b) proton number, (c) neutron number, (d) electron number.

6. How many naturally occurring elements are there? *(a) 90, (b) 92, (c) 103, (d) 106.

7. A period in the periodic table contains elements with (a) chemical symbols beginning with the same letter, (b) similar chemical properties, *(c) consecutively increasing atomic numbers, (d) the same atomic masses.

8. Which of the following is a transuranic element? (a) Ru, *(b) Pu, (c) Pa, (d) Nb.

9. The strongest chemical bond generally occurs in a(n) *(a) ionic bond, (b) covalent bond, (c) polar bond, (d) U.S. savings bond.

10. An antiatom of the most common atom of uranium would have in its nucleus *(a) 92 antiprotons, (b) 92 positrons, (c) 143 antineutrons, (d) 146 neutrons.

11. An early supporter of the theory that matter consisted of discrete particles (atoms) was (a) Plato, (b) Aristotle, *(c) Democritus, (d) Virgil.

12. Early direct evidence for the atomic theory was furnished by *(a) Brownian motion, (b) antiparticles, (c) states of matter, (d) the periodic table.

13. The diameter of an atom is of the order of (a) 10^{-5} cm, *(b) 10^{-8} cm, (c) 10^{-10} cm, (d) 10^{-8} $\overset{\text{o}}{\text{A}}$

14. In the over-simplified solar system model of the atom, the "planets" correspond to (a) protons, *(b) electrons, (c) neutrons, (d) the neucleus.

15. An atom of particular species or element is defined by the number of (a) antiparticles, (b) electrons, (c) neutrons, *(d) protons.

16. The elements in the periodic table are arranged horizontally in rows with successively increasing (a) atomic masses, *(b) atomic numbers, (c) melting points, (d) isotopes.

17. Elements with similar chemical properties in the periodic table form a (a) period, (b) family of siotopes, *(c) vertical column, (d) diagonal line.

18. Chemical bonds associated with substances having high melting points are (a) covalent bonds, *(b) ionic bonds, (c) polar bonds, (d) Brownian bonds.

19. The phase of matter that has a definite volume but no definite shape is *(a) liquid, (b) solid, (c) gas, (d) plasma.

20. Particles and their antiparticles (a) form antiatoms, (b) make up a plasma, (c) result from covalent bonding, *(d) annihilate each other.

21. The smallest unit of an element that can exist alone or in combination with other atoms is called a/an (a) electron, (b) proton, *(c) atom, (d) neutron.

22. A substance in which all the atoms have the same number of protons is called a/an (a) compound, *(b) element, (c) molecule, (d) none of these.

23. An atom or molecule with a net electrical charge due to the transfer (loss or gain) of one or more electrons is known as a/an (a) isotope, (b) polar atom or molecule, *(c) ion, (d) none of these.

24. Atoms or nuclei of the same species (same number of nuclear protons) having different numbers of neutrons are called (a) molecules, (b) compounds, (c) ions, *(d) isotopes.

25. A horizontal row in the periodic table is called a (a) group, *(b) period, (c) family, (d) none of these.

26. A _____ is a group of two or more atoms held together by forces called chemical bonds. *(a) molecule, (b) compound, (c) period, (d) family.

27. A/an _____ is a pure substance made up of one or more elements. (a) molecule, *(b) compound, (c) isotope, (d) ion.

Chapter 7

28. A _____ bond is formed by sharing electrons
between atoms. (a) neutral, (b) polar, (c)
non-polar, *(d) covalent

29. A covalent bond in which there is an unequal sharing
of electrons between atoms such that there is an
unsymmetric charge distribution or molecular regions
of net charge is known as a _____ bond.
(a) non-polar, *(b) polar, (c) neutral, (d) none of
these.

30. Which of the following is not a phase of matter?
(a) liquid, (b) gas, *(c) ion, (d) plasma.

Completion

1. The atomic nucleus is made up of protons and
neutrons.

2. The phase of matter which has a definite volume but
no definite shape is the liquid phase.

3. An antiproton has a negative charge.

4. The subatomic particle which has the largest mass is
the neutron.

5. The characteristic which distinguishes the atoms of
one element from the atoms of another is the number
of protons.

6. The periodic table arranges elements in a vertical
columns that have similar chemical properties.

7. A great many elements were discovered when
alchemists tried to produce gold from other metals.

8. A phenomenon which indicates the existance of
molecules is Brownian motion.

9. The two physical properties which determine the
phase of a substance are temperature and pressure.

10. Molecules such as N_2 and O_2 are called diatomic.

11. Dalton developed explanations of several laws of
chemistry using atomic theory.

12. The Brownian motion of pollen grains suspended in
water is the result of a barrage of randomly moving
water molecules.

13. Most of the volume of an atom is <u>empty space</u>.

14. One atom has 6 protons and 6 neutrons, and another atom has 6 protons and 7 neutrons. Both are <u>isotopes</u> of carbon.

15. The periodic table of elements was developed by <u>Mendeleev</u>.

16. Currently, there are <u>90</u> naturally occurring elements.

17. Elements with atomic numbers greater than 92 are called <u>transuranic (radioactive)</u> elements.

18. Ionic bonding involves a <u>transfer or sharing</u> of electrons.

19. A <u>gas</u> has no definite shape or volume.

20. Particle and antiparticle annihilation involves the conversion of <u>mass</u> to <u>energy</u>.

21. The <u>atom</u> is the smallest unit of an element that can exist alone or in combination with other atoms.

22. An <u>element</u> is a substance in which all the atoms have the same number of protons.

23. The erratic motion of small particles in suspension due to collisions with the molecules of the suspension medium is known as <u>Brownian</u> motion.

24. An <u>ion</u> is an atom or molecule with a net electrical charge due to the transfer (loss or gain) of one or more electrons.

25. <u>Isotopes</u> are atoms or nuclei of the same species (same number of nuclear protons) having different numbers of neutrons.

26. A horizontal row in the periodic table is called a <u>period</u>.

27. A <u>molecule</u> is a group of two or more atoms held together by forces called chemical bonds.

28. A <u>compound</u> is a pure substance made up of one or more elements.

29. A bond formed by a transfer of electrons between atoms is called a/an <u>ionic</u> bond.

30. A bond formed by a sharing of electrons between atoms is called a/an <u>covalent</u> bond.

31. A <u>polar</u> bond is a covalent bond in which there is an <u>unequal</u> sharing of electrons between atoms such that there is an unsymmetric charge distribution or molecular regions of net charge.

32. The four phases of matter are solid, liquid, gas, and <u>plasma</u>.

33. The charge of an antiproton is <u>negative</u>.

34. Electrons traveling around a nucleus can have only definite amounts of <u>energy</u>.

35. All atoms which have a proton number greater than <u>92</u> are transuranic.

36. The central force of an atomic electron is supplied chiefly by the <u>electrical</u> force between the electron and the nucleus (protons).

37. All elements having proton numbers greater than <u>83</u> are radioactive.

Matching

(Choose the appropriate answer from the list on the right.)

d	1. atom	a.	positively charge sub-atomic particle
j	2. compound	b.	an atom having an unequal number of electron and protons
g	3. covalent bond	c.	horizonal row
k	4. electron	d.	the major particle of an element
o	5. element	e.	contains protons and neutrons
b	6. ion	f.	a group of two or more atoms held together by chemical bonds
p	7. isotope	g.	electrons are mutually shared rather than transferred
f	8. molecule	h.	nonsymmetric distribution of charge
e	9. nucleus	i.	A "gas" of charged particles
i	10. plasma	j.	a substance made up of more than one element
h	11. polar	k.	the subatomic particle with the smallest mass
a	12. proton	l.	Mendeleev
l	13. periodic table	m.	electron transfer
m	14. ionic bond	n.	similar chemical properties
n	15. group	o.	same proton number
c	16. period	p.	same proton number different neutron numbers

Chapter 8   Solids

Answers to Questions

1.   Connecting springs at random angles.

2.   (a)   Ions at the centers of faces of cube,
      (b)   Ion at center of cube.

3.   Grain direction is along direction of polymer
      chains.

4.   Different atomic structures.

5.   Atoms in uranium not as compact as atoms in gold.

6.   Yes, more mass per volume.

7.   Mercury contracts on freezing.

8.   Because of dissolved salt (NaCl) and other ions and
      elements that are more dense than water.

9. $V = 5 \times 10 \times 20 = 1000$ cm^3 (and densities from Table
      8.1)

      (a) m = $\rho_{Au} V$ = $(19.3)(10^3)$ = 19.3 kg = 19.3 $\times 10^3$ g

           m = $\rho_{Pb} V$ = $(11.5)(10^3)$ = 11.5 kg = 11.5 $\times 10^3$ g

      (b)   Au: 19.3 kg (2.2 lb/kg) = 42.5 lb

            Pb: 11.5 kg (2.2 lb/kg) = 25.3 lb

10.   The kilogram was originally defined as the mass of
      water in a cubic deciliter and not cubic meter.

11.   Density

12. A_2 = area of square = 4 x 4 = 16 cm², and A_1 =
      area of circle = πr² = π(1)² = π cm² = 3.14

      cm², and $4A_1$ = 4π = 12.6 cm². And,

      A_3 = area of empty space = A_2 - A_1 = 16 - 12.6 =

      3.4 cm². Then, A_3/A_2 (x 100%) = 3.4/16 (x 100%) =

      21.3%

13.   Crystalline and high melting points, but different
      bonding.

14.   The stronger the bonding, the higher the melting
      point.

15. Weak bonding allows sublimation at normal temperatures.

16. Sublimation, particularly in frost-free freezers with circulating air.

17. Open lattice structure, which begins to form below 4°C as implied by the decreasing density.

18. They do not decompose and remain indefinitely. Unstablizers would promote breakdown.

19. Pretty bare.

20. Electrons less tightly bound.

21. Different bondings.

22. The sandwiched copper is not uniformly dispersed, so not a true alloy, although individual parts may be alloys.

23. Because the density of the zinc component in brass is less than that of copper (Table 10.1).

24. The same length, since the spring constant depends on <u>change</u> in length and not on the length of the spring.

25. One centimeter each, since k's of springs are the same.

26. They vary in different directions.

27. Wood is cheaper and more elastic. Rubber or wooden mallets are used in some cases to prevent permanent deformation of the object.

28. More elastic so as to absorb shocks and cheaper.

29. Iridium   Lead
    Diamond   Aluminum

30. Ceramic materials fracture without plastic deformation, but are more dense (heavier).

31. The breaking of the material is a function of: (a) the type of material (b) the width and thickness of the material (c) the distance between the supports (d) applying the impact force to a small area so as to produce a large force per unit area (e) timing the impact so that maximum momentum and energy are are transferred to the material.

Chapter 8

SAMPLE TEST QUESTIONS

Multiple Choice

1. An amorphous solid has (a) a definite melting point
    temperature, (b) an orderly array of particles, (c)
    an X-ray diffraction pattern, *(d) a random particle
    arrangement.

2. Which of the following is an amorphous solid? (a)
    table salt, *(b) glass, (c) diamond, (d) zinc.

3. The material with the greatest density is (a) water,
    (b) zinc, *(c) osmium, (d) iron.

4. The type of solid that generally has a relatively
    low melting point is (a) ionic, *(b) molecular, (c)
    macromolecular, (d) metallic.

5. Which type of solid has interparticle covalent
    bonding? (a) ionic, *(b) molecular, (c)
    macromolecular, (d) metallic.

6. A polymer substance is (a) always a plastic, (b) one
    giant molecule, (c) always synthetic, *(d) made up
    of monomers.

7. Alloys (a) have only metallic elements, (b) consist
    of only two elements, *(c) have properties different
    than those of the individual elements, (d) are poor
    electrical conductors.

8. To some degree, all solids have *(a) elasticity, (b)
    plasticity, (c) ductility, (d) malleability.

9. One of the good mechanical properties of lead is (a)
    hardness, (b) ductility, *(c) malleability, (d)
    brittleness.

10. A ceramic material exhibits a great deal of (a)
    plasticity, (b) ductility, (c) malleability,
    *(d) brittleness.

11. A solid with a completely random particle
    arrangement is said to be (a) crystalline, *(b)
    amorphous, (c) plastic, (d) a liquid crystal.

12. Which type of solids have relatively high and
    well-defined melting points? *(a) ionic, (b)
    molecular, (c) amorphous, (d) plastics.

13. Which of the following is a hard, nonbrittle substance? (a) lead, (b) graphite, (c) quartz, *(d) diamond.

14. Which type of solid can be cleaved? (a) amorphous, (b) polymer, *(c) crystalline, (d) molecular.

15. NaCl (sodium chloride), which is common table salt, (a) is a molecular solid, (b) is used in photochromic glasses, *(c) has a simple cubic lattice structure, (d) has an amorphous structure.

16. The density of water is generally given as

    (a) 1 kg/m^3, (b) 100 kg/m^3, (c) 100 g/cm^3,

    *(d) 1 g/cm^3.

17. Water has its maximum density at (a) $0^{\circ}C$, *(b) $4^{\circ}C$, (c) 10° C, (d) $100^{\circ}C$.

18. A plastic is *(a) made up of monomers, (b) a macro-molecular solid, (c) a crystalline solid, (d) an alloy.

19. The alloy bronze is made up of (a) iron and carbon, (b) copper and zinc, *(c) copper and tin, (d) chromium and nickel.

20. Plastic deformation occurs (a) chiefly in ceramic materials, (b) only for metals, (c) only in plastics, *(d) when the elastic limit is reached.

21. The fundamental repeating units in large molecules are called (a) polymers, *(b) monomers, (c) macromolecules, (d) ionics.

22. A solid with a regular or orderly particle arrangement is a lattice structure known as a/an _____ solid. (a) amorphous, (b) ionic *(c) crystalline, (d) covalent.

23. The compactness of the particles in a material is measured in units of (a) g/m, (b) g/m², *(c) kg/m^3, (d) none of these.

24. _____ is a solid consisting of oppositely charged ions. (a) ice, *(b) salt, (c) diamond, (d) silicon dioxide.

25. A/an _____ solid consists of covalent bonded atoms such that in effect the solid consists of one large molecule. (a) micromolecular, *(b) macromolecular, (c) amorphous, (d) ionic.

26. A solid consisting of a lattice of positive ions surrounded by a "sea" of electrons, which gives rise to good electrical conduction is known as a/an *(a) metalic solid, (b) ionic solid, (c) alloy, (d) polymer.

27. In applying a force to a solid it is common to speak of applying a stress, where stress is defined as the (a) applied force, (b) force x area, *(c) force/area, (d) force/volume.

28. Hooke's law states the relationship between (a) force and area, (b) stress and force, (c) force and strain, *(d) stress and strain.

29. The geometric pattern or arrangement of particles in a crystalline solid is known as the (a) ionic structure, *(b) lattice, (c) macromolecular pattern, (d) none of these.

30. An in-between state in which a substance with liquid properties shows some degree of molecular order as in a crystalline solid as known as a/an *(a) liquid crystal, (b) macromolecular solid, (c) micromolecular solid, (d) amorphous solid.

Completion

1. When water solidifies into ice, the volume increases and the density decreases.

    it expands

2. The most dense solid on Earth is osmium.

3. A polymer substance that can flow under heat and pressure, and can be molded into various shapes is plastic.

4. The condition at which a substance is on the verge of becoming permanently deformed is the elastic limit.

5. Within certain limits for springs, the force is directly proportional to the displacement.

6. Diamond is one of the hardest crystalline substances.

7. An object's density is less than that of water. When the object is placed in water, the object will float.

8. A common alloy made of copper and tin is bronze.

9. Some metals become harder and more brittle as a result of work hardening.

10. Glass is an amorphous solid.

11. If a crystalline solid is not cleaved along a cleavage plane, it will fracture.

12. Graphite, a form of carbon, is used in "lead" pencils.

13. Photochromic glasses darken when exposed to ultraviolet radiation.

14. LCD stands for Liquid Crystal Display.

15. The densest solid is the metal osmium.

16. The SI unit of density is Kg/m^3.

17. Ice has an open, hexagonal (six-sided) lattice pattern.

18. A plastic polymer will flow under heat and pressure and can be molded into different shapes.

19. Brass is an alloy of copper and zinc.

20. Plastic deformation is permanent.

21. The fundamental repeating units in large molecules are called monomers.

22. Plasticity is that property of a solid characterized by permanent deformation when a force or stress is removed.

23. A solid with a regular or orderly perticle arrangement in a lattice structure is known as a crystalline solid.

24. Lattice is the geometric pattern or arrangement of particles in a crystalline solid.

25. A/an ionic solid (a compound) consists of oppositely charged ions.

26. The ratio of stress to strain is called <u>Hooke's</u> law.

27. Stress has the units of <u>force</u> per unit area.

28. A/an <u>alloy</u> is a blend of two or more metallic elements or metallic and nonmetallic elements.

29. The property of a solid whereby it returns to its original shape after a distorting force or stress is removed is called <u>elasticity</u>.

30. A/an <u>amorphous</u> solid consists of random particle arrangement.

31. A solid consisting of covalently bonded atoms, such that in effect the solid consists of one huge molecule is called a/an <u>macromolecular</u> solid.

32. Because of the strong ionic bonding, ionic solids have relatively <u>high</u> melting points.

33. A liquid with some degree of molecular order is called a/an <u>liquid crystal</u>.

34. Ionic solids have bonds that result from a <u>transfer</u> of electrons.

35. Plastics are made primarily from <u>petroleum</u>.

Matching

(Choose the appropriate answer from the list on the right.)

_k_ 1. solids with a regular arrangement of particles

a. liquid crystal

_s_ 2. change of phase from solid directly to gas

b. polymer

_j_ 3. solids with random molecular arrangement

c. orderly array

_o_ 4. the small molecular unit which forms giant molecules

d. ionic solid

_q_ 5. fractional change in an object's dimension

e. covalent bonding

_r_ 6. force per unit area

f. snowflakes

_n_ 7. reflects the internal resistance of a solid to having its particles forced closer together.

g. brass

h. F proportional to x

_m_ 8. return to original shape when force is removed.

i. photochromic glasses

_p_ 9. plastic deformation

_l_ 10. mass per unit volume

j. amorphous

_a_ 11. in-between state

k. crystalline

_d_ 12. oppositely charge particles

l. density

m. elsticity

_t_ 13. one huge molecule

n. hardness

_b_ 14. repeating monomer units

o. monomer

_c_ 15. lattice

p. plasticity

_i_ 16. AgCl

q. strain

_f_ 17. hexagonal

r. stress

_g_ 18. alloy

s. sublimation

_h_ 19. Hooke's law

t. macromolecule

_e_ 20. Sharing of electrons

Chapter 9   Liquids

Answers to Questions

1. Six (4 sides, 2 ends) if unsupported. (See Question 2).
   Greater pressure on end (smaller area). Force same in all instances, i.e., weight of brick.

2. Three (long edge, short edge, and corner).
   Greatest pressure on corner.

3. Yes, since $\rho g = mg/V = w/V = D$.

4. Lower hole, greater pressure. Same rates if lower hole increased to appropriate size.

5. Same, since same liquid heights, $p = \rho gh$.

6. Greater pressure at base. Pressure depends on water depth of application.

7. Use pressure-depth relation with density
   ($\rho = 10^3$ kg/m^3 and g = 10 m/s²).

   $p = \rho gh = (10^3)(10)(1) = 10^4$ N/m² = 10 kPa

   Total pressure = 2000 + 10 = 2010 kPa

8. Yes, greater pressure (Pascal's principle).

9. Force multiplication and corresponding distance reduction.

10. 50 cm/100 = 0.50 cm.

11. $P_i = F/(100\ A) = P_o = (F/100)/A$, so MA = 1/100 , and
    $d_i/d_o = 1/100$ or $d_i = 10$ cm (1/100) = 0.10 cm

12. Overall density of ship less than that of sea water.
    $V = w/D = 6 \times 10^8/10^4 = 6 \times 10^4$ m^3.

13. Higher. Sea water has greater density and less water needs to be displaced.

14. Overall human density is less than that of water.
    Density changed by inhaling and exhaling.

15. Density of salt water greater than that of fresh water and objects (humans) float higher.

16. Kinetic energy due to work done by buoyant force.

76

17. Liquid (mercury) density greater than density of iron.

18. Ice cubes higher if density of drink is less than of water, and would be if alcohol content sufficient. (Some drinks have other additivies.) Party float lower since fruit generally increases density.

19. Varying density by adding water to jug.

20. Attach sinker by string. Measurements with sinker submerged only, and sinker plus object to get object's volume. Mass or weight determined with scales.

21. (a) Smaller above 1.000 mark. Hydrometer sinks more in less dense liquids. (b) Specific gravity depends on concentration of acid or antifreeze.

22. No, same displacement, same buoyant force.

23. Yes, increase since net downward weight increases (action-reaction pairs of Newton's third law).

24. To decrease overall (average) density.

25. Overall (average) density of boat (includes contents) increases.

    Bailing helps to stablize or reduce density if bailing rate is equal to or greater than water input rate.

26. Due to surface tension.

27. Due to surface tension. Distortion due to gravity.

28. The soap-water solution has greater molecular interactions and wetting. Other additives, e.g., glycerol make bubbles longer lasting.

29. Mercury has a high surface tension (over 6 times greater than water) due to large cohesive forces and so forms spherical drops while not wetting surfaces.

30. PE decrease goes into pressure and flow speed increase. (Consider water towers on hllls.)

31. Constrict pipe to decrease pressure and increase flow speed.

32. Molecular interactions.

Chapter 9

33. No, surface tension and action would be decreased.

34. Cream has greater viscosity. Milk has greater density since cream floats on milk.

35. Large viscosity, poor lubrication, and car probably wouldn't start due to thick oil on very cold days.

SAMPLE TEST QUESTIONS

Multiple Choice

1. A fluid (a) has a definite shape, (b) is always incompressible, *(c) cannot support a shear stress, (d) refers to liquids only.

2. Which of the following units expresses pressure?

   (a) Pa/m^2, *(b) $kg/m\text{-}s^2$, (c) $N\text{-}m^2$, (d) $N\text{-}s/m^2$

3. The pressure in a liquid depends on (a) the weight density of the liquid, (b) the depth in the liquid, (c) the mass density of the liquid, *(d) all of the preceding.

4. A hydraulic jack is an application of (a) Archimedes' principle, (b) Bernoulli's principle, *(c) Pascal's principle, (d) Newton's principle.

5. The buoyant force on a submerged object (a) depends on the weight of the object, (b) is greater in a less dense liquid, (c) increases with depth in an incompressible liquid, *(d) depends on the volume of liquid displaced.

6. If the density of a block of some material is 0.80 g/cm^3, when placed in water (a) it will sink, (b) it will sink only if the water wets its surface, *(c) 20 percent of its volume will be above the surface of the water, (d) 80 percent of its volume will be above the surface of the water.

7. Capillary action depends on (a) Pascal's principle, (b) buoyancy, *(c) molecular interactions, (d) viscosity.

8. In streamline flow, (a) particles cross paths, *(b) the fluid speed increases when the streamlines are closer together, (c) there can be small whirlpools and eddies, (d) the streamlines cross each other.

9. A fluid flows in a horizontal pipe. If the cross section of the pipe becomes smaller, (a) the pressure increases, *(b) the speed increases, (c) both the pressure and speed increase, (d) neither the pressure or speed changes.

10. The viscosity of motor oil *(a) is greater for SAE 30 than for SAE 10, (b) increases with temperature, (c) remains the same from start up to normal running, (d) requires that SAE 40 be used in winter.

11. Which of the following is not a fluid at room temperature and atmospheric pressure? (a) mercury, (b) oxygen, (c) carbon dioxide, *(d) carbon.

12. In a full, closed container of a liquid, the pressure does not depend on *(a) the shape of the container, (b) the liquid's mass density, (c) the acceleration due to gravity, (d) the depth.

13. A mechanical advantage can be obtained by the application of (a) Archimedes' principle, *(b) Pascal's principle, (c) Bernoulli's principle, (d) surface tension.

14. The buoyant force is described by (a) Bernoulli's principle, (b) streamline flow, *(c) Archimedes' principle, (d) Pascal's principle.

15. Surface tension is a result of (a) Bernoulli's principle, (b) viscosity, *(c) intermolecular forces, (d) buoyancy.

16. Cleansing action is increased by a reduction of (a) buoyance, (b) capillary action, (c) viscosity, *(d) surface tension.

17. Streamline flow is (a) described by Archimedes' principle, *(b) ideal, (c) turbulent, (d) a pressure-depth relation.

18. Work-energy conservation in ideal fluid flow is expressed in (a) Archemedes' principle, (b) SAE number, (c) Pascal's principle, *(d) Bernoulli's principle.

19. Viscosity *(a) decreases with increasing temperature, (b) affects the pressure-depth relationship, (c) is a factor in buoyancy, (d) causes surface tension.

20. The surface tension of water (a) is expressed in terms of SAE number, (b) affects buoyancy, *(c) decreases with increasing temperature, (d) increases when a detergent is added.

21. Liquids (a) have a definite volue, (b) have no definite shape, (c) are essentially incompressible, *(d) all of the above.

22. Pressure is defined as (a) force x area, (b) force/area², *(c) force/area, (d) force/volume

23. Pressure applied to an enclosed liquid (fluid) is transmitted undiminished to every other part of the liquid and to the walls of its container. This is a statement of _____ principle. (a) Archimedes', *(b) Pascal's , (c) Bernoulli's, (d) Newton's.

24. Archimede's principle applies to (a) capillary action, (b) surface tension, (c) streamline flow, *(d) bouyancy.

25. The crescent-shaped surface of a liquid is called the (a) curve, (b) semicircle, *(c) meniscus, (d) lens.

26. Capillary action is a function of (a) cohesive forces, *(b) adhesive forces and surface tension, (c) molecular potential energy, (d) none of these.

27. Viscoscity refers to _____ of a fluid to flow. (a) pressure, (b) streamline action, *(c) internal resistance, (d) work-energy.

28. A fluid is a substance that (a) is incompressible, *(b) can flow, (c) can support a shearing stress, (d) cannot exist as a gas.

29. The rising of a hot air balloon is explained by *(a) Archimedes' principle, (b) Pascal's principle, (c) Bernoulli's principle, (d) none of these.

30. Detergents (a) have no effect on surface tension, (b) increase surface tension, *(c) decrease surface tension (d) none of these.

Completion

1. That pressure applied to an enclosed fluid is transmitted undiminished to every other part of the fluid is known as Pascal's principle.

2. According to Archimedes' principle, an object in a fluid is buoyed up by a force equal to the weight of the volume of the fluid it displaces.

3. If the density of an object is greater than the density of a liquid, the object will sink when placed in the liquid.

4. An object is "wet" when placed in water because the adhesive forces are greater than the cohesive forces.

5. Bernoulli's principle reflects the conservation of energy.

6. The streamlines in streamline flow never cross or intersect.

7. Gases and liquids are classified as fluids.

8. A basic difference between solids and fluids is that fluids cannot support a shear force.

9. If an object has the same density as that of a liquid, the object will remain at any level when submersed in the liquid.

10. A liquid has a definite volume but no definite shape.

11. The pressure in a liquid depends on the liquid's (weight) density.

12. According to Pascal's principle, pressure is transmitted undiminished in a fluid.

13. If the force is doubled and the area of application is reduced by one-half, the pressure increases by a factor of four.

14. When an object displaces more liquid, the buoyant force increases.

15. A submarine submerges by increasing its density (flooding ballast tanks).

16. Water bugs can walk on water because of <u>surface tension</u>.

17. Detergents <u>reduce</u> surface tension.

18. When the cohesive forces exceed the adhesive forces for a liquid in a tube, the surface of the liquid has a <u>convex</u> meniscus.

19. When streamlines cross each other, the fluid flow becomes <u>turbulent</u>.

20. The internal resistance of a fluid is expressed in terms of <u>viscosity</u>.

21. Liquids have definite volume, but no definite <u>shape</u>.

22. A fluid is a substance that can <u>flow</u>.

23. Pressure is defined as force per unit <u>area</u>.

24. The SI units for pressure are <u>N/m² or pascals</u>.

25. The liquid pressure on the bottom of a container depends on the <u>height</u> of the liquid and its <u>weight density</u>.

26. Pascal's principle states that in a confined liquid, the pressure is transmitted <u>undiminished</u> to all walls of its containers.

27. Archimede's principle states that an object wholly or partly immersed in a fluid experiences a <u>bouyant</u> force equal to the <u>weight</u> of the fluid displaced.

28. Surface tension is defined as the <u>force</u> that causes the surface of a liquid to contract.

29. Detergents <u>reduce</u> surface tension.

30. Capillary action is due to forces of adhesion and <u>surface tension</u>.

31. Bernoulli's principle states that when a fluid moving in streamline flow increases in velocity, the pressure will <u>decrease</u>.

32. The <u>internal resistance</u> of a fluid to flow is known as viscosity.

33. The level of mercury in a capillary tube is <u>depressed</u>.

34. Surface tension causes free liquid droplets to have <u>spherical</u> shapes.

35. The cohesive forces are <u>less</u> than the adhesive forces in a substance that wets another substance.

36. An object is placed 2 m deep in a pool. When placed 4 m deep, the pressure due to the water is <u>doubled</u>.

37. The hydraulic jack is an application of <u>Pascal's</u> principle.

38. The viscosity of SAE 10W-40 motor oil is <u>greater</u> at winter temperatures than at summer temperatures.

Chapter 9

Matching

(Choose the appropriate answer from the list on the right.)

k	1.	liquid	a.	force/area
e	2.	fluid	b.	SAE 10
d	3.	Archimede's principle	c.	spherical droplets
j	4.	Pascal's principle	d.	bouyant force
i	5.	Bernoulli's principle	e.	gas or liquid
p	6.	adhesion	f.	cohesion and adhesion
c	7.	surface tension	g.	internal resistance
g	8.	viscosity	h.	attraction between like molecules
f	9.	capillary action	i.	work and energy
h	10.	cohesion	j.	hydraulic jack
m	11.	streamline flow	k.	definite volume and shape
n	12.	pressure in a liquid	l.	crescent-shaped
b	13.	winter motor oil	m.	parallel paths
o	14.	summer motor oil	n.	a function of height
a	15.	pressure	o.	SAE 40
l	16.	meniscus	p.	attraction between unlike molecules

Chapter 10   Gases

Answers to Questions

1. (a) Volume decreases, (b) temperature decreases.

2. (a) Density doubles, (b) pressure doubles. Work is done in compressing gas.

3. Work done in compressing gas causes temperature increase.

   Increased pressure in tire makes it more difficult to add air.

4. Pressure (-volume) increase.

5. The final draining of liquid increases the volume of the trapped air and its pressure is reduced. The pressure difference between this and atomspheric pressure prevents further draining.

6. Atmospheric pressure forces air in due to pressure difference. When pouring slowly with tilt, pressure in bottle is equalized.

7. Until the weight of water in glass equals the upward force due to pressure difference.

8. An air vent to prevent pressure difference that would retard liquid flow.

9. Atmospheric pressure is less and less "rising" ingredient is needed.

10. Smaller tire has less contact area and greater load pressure, so greater internal pressure is needed.

11. (a) In filling, air is expelled from bulb and pressure difference fills tube. In dispensing, pressure on bulb greater than atmospheric pressure. (b) Air is injected to create pressure difference for filling syringe. If not done, pressures in bottle and syringe equalize and no liquid flows.

12. Out the window due to pressure difference.

13. Wetted glass lip and card forms a seal and atmospheric pressure holds water in glass. Same effect horizontally.

14. Forcing out of air in attaching suction cup gives pressure difference which causes it to "stick." Would not work on moon -- no atmosphere.

15. "Pushed", since pressure in plunger is reduced.

16. (a) 60 in., (b) 15 in.

17. Not with the help of atmospheric pressure. No atmosphere.

18. Death Valley, greater atmospheric pressure.

19. Because atmospheric pressure will support a limited column (10.3 m).

20. Atmospheric pressure gets less and internal pressure causes volume increase. As air becomes less dense, buoyant force decreases until buoyant force and balloon weight are equal.

21. As air in balloon cools, it becomes more dense and buoyancy decreases.

22. The density of air decreases with altitude, and so does the buoyancy. At some altitude the weight of the balloon is equal to the buoyant force.

23. So as to displace a large volume of low-density air to have an adequate buoyant force. Balloons have no horizontal guidance system, whereas blimps have propellers and can navigate.

24. Downward stroke lets water through upper valve and is lifted on upstroke. Upstroke causes pressure reduction in chamber and pressure difference fills chamber. When not used for long period, pump gasket becomes dry and chamber is not airtight.

25. No, pressure difference causes air to be forced in.

26. Pulse rate is number of heart beats per minute. Blood pressure of 120/80 is systolic pressure/distolic pressure (occurs with each beat cycle).

27. Lung pressure.

28. Slower speed, greater pressure (Bernoulli's principle).

29. Greater weight requires greater lift force, which requires greater (air) speeds and so longer runways are needed to get up speed. Similarly in landing.

30. (a) Flapping wings and action-reaction with air.
    (b) Conservation of momentum.

31. Slight lift wings curved, but primarily fall
    retarded by air resistance.

32. Reaction lift force due to air forced downward.

33. Rotational motion gives lift force due to air motion
    over curved surface (and lateral curving force
    similar to baseball), also air resistance effects.
    When rotation slows, Frisbee falls.

34. Reduced air pressure over curved top.

35. Greater air speed between vehicles gives reduced
    pressure (Bernoulli effect), and inward force due to
    pressure difference.

36. Air flow over tube reduces pressure and rising
    liquid is "atomized" by air flow.

37. No, heated air from room is forced up chimney.

38. Yes, downward.

39. About 10 - 12 breaths per minute. Taking 10 breaths
    per minute, then 5 L/min (1 min/10 breaths) = 0.5
    L/breath.

40. 6 x 4 = 24 in²; 14.7 lb/in² x 24 in² = 352.8 lb.

41. 1 atm = 100 kPa = 10^5 N/m², and A = 0.20 m².

    F = pA = $(10^5)(0.20)$ = 2 x 10^4 N (=4500 lb)

42. A = 1.0 m², p = 1 atm = 10^5 N/m²

    (a) F = pA = $(10^5)(1.0)$ = 1.0 x 10^5 N

    (b) 1.0 x 10^5 N (0.225 lb/1 N) = 2.25 x 10^4 lb =
        22,500 lb. The table can be picked up because
        there is a nearly equal upward pressure on
        bottom of table.

Chapter 10

SAMPLE TEST QUESTIONS

Multiple Choice

1. At a constant temperature, when the pressure of a quantity of gas decreases, the volume (a) increases, *(b) decreases, (c) remains the same, (d) is not predicted the perfect gas law.

2. When energy (heat) is added to a quantity of gas, (a) the pressure always increases, (b) the volume always increases, (c) the volume always decreases, *(d) the pressure or volume or both increases.

3. The pressure gauge on a rigid container with a gas is noted to decrease. This could be due to (a) gas being added to the container, (b) a volume decrease, *(c) a temperature decrease, (d) a change in change in atmospheric pressure.

4. Only about 1 percent of the atmospheric gases lie above an altitude of (a) 10 km, (b) 20 km, *(c) 30 km, (d) 50 km.

5. The SI unit of atmospheric pressure is (a) mm Hg, *(b) pascal, (c) torr, (d) millibar.

6. Which one of the following is not the approximate normal pressure of one atmosphere at sea level? *(a) 100 Pa, (b) 760 torr, (c) 1000 mb, (d) 100 kPa.

7. The barometer is an instrument used to measure (a) volume, (b) temperature, *(c) pressure, (d) fluid flow.

8. In a piston-cylinder force pump, a partial vacuum is created in the cylinder (a) during the forward stroke, *(b) during the back stroke, (c) during both strokes, (d) during neither stroke.

9. A normal systolic blood pressure would be (a) 70 torr, (b) 80 torr, (c) 125 torr, *(d) 180 torr.

10. A spinning baseball curves *(a) toward the high-velocity side, (b) toward the high-pressure side, (c) in a nonviscous fluid, (d) because of atmospheric pressure.

11. The perfect gas law does not depend on (a) pressure, (b) volume, *(c) gravity, (d) density.

When the volume of a gas is decreased, (a) the pressure must increase, *(b) the density must increase, (c) the temperature must increase, (d) Boltzmann's constant changes.

13. Normal atmospheric pressure is about (a) 30 lb/in², (b) 76 torr, (c) 100 mb, *(d) 100 kPa.

14. Which one of the following is not used to measure pressure? (a) sphygmomanometer, (b) aneorid barometer, (c) altimeter, *(d) syphon.

15. The barometer was invented by (a) Galileo, (b) Bernoulli, *(c) Torricelli, (d) Magnus.

16. When the burner of a hot-air balloon is out of propane, the flight could be extended by *(a) throwing the burner overboard, (b) letting some hot air out of the balloon, (c) jumping up and down, (d) shifting the gondola's contents to one side.

17. A force pump takes liquid into the cylinder (a) on the forward stroke, *(b) on the back stroke, (c) on both strokes, (d) when the working volume of the cylinder is decreased.

18. The "relaxation" pressure of the circulatory system is called (a) barometric pressure, *(b) diastolic pressure, (c) systolic pressure, (d) air pressure.

19. The heart is effectively a *(a) force pump, (b) lift pump, (c) barometer, (d) vacuum cleaner.

20. The temperature in the perfect gas law must be expressed in (a) degrees Fahrenheit, (b) degrees Celsius, *(c) kelvins, (d) any of the preceding.

21. Gases (a) have a definite volume, (b) have a definite shape, (c) have no internal energy, *(d) are fluids.

22. The pressure of a gas depends on its (a) volume, (b) temperature, (c) internal kinetic energy, *(d) all of these.

23. Boyle's gas law states the relationship between pressure and volume of a gas at *(a) constant temperature, (b) varying temperature, (c) constant heat, (d) none of these.

24. The ratio of nitrogen to oxygen in the Earth's atmosphere is approximately (a) 2 to 1, (b) 3 to 1, *(c) 4 to 1, (d) none of these.

25. The pressure times the volume of a perfect gas is proportional to the _____ of the gas. (a) absolute temperature, (b) number of molecules, (c) I/V, *(d) all of these.

26. Atmosphere pressure is measured in units of (a) pounds per square inch, (b) millimeters of mercury, (c) pascals, *(d) all of these.

27. Bernoulli's principle states the relationship between _____ and _____ in streamline flow of a fluid. (a) pressure and temperature, *(b) pressure and speed, (c) pressure and volume, (d) none of these.

28. The perfect gas law states that pressure of a gas is (a) proportional to the volume, (b) indirectly proportional to the temperature, *(c) directly proportional to the temperature, (d) inversely proportional to the number of particles.

29. Which of the following always increases in direct proportion, when the temperature of a perfect gas is increased? (a) pressure, (b) volume, *(c) kinetic energy, (d) number of particles.

30. The units used for values of pressure, volume, and temperature in the perfect gas law must be in _____ units. *(a) SI, (b) customary, (c) British, (d) none of these.

Completion

1. Gases have no definite shapes because there are very weak <u>cohesive</u> forces between the gas molecules.

2. The kinetic energy of the gas molecules of a perfect gas is its <u>internal total</u> energy.

3. The Earth's atmosphere is more dense near the surface of the Earth because of <u>gravity or gravitational attraction</u>.

4. One atmosphere is equivalent to <u>760</u> torr.

   For 5, 6, and 7: A perfect gas has an original pressure p_o, and volume V_o, and temperature T_o (for each question.

5. If the volume is constant and the pressure changes to $p_o/2$, the temperature is then <u>$T_o/2$</u>

6. If the temperature is constant and the pressure is increased to $2p_o$, the volume is <u>$V_o/2$</u>.

7. If the volume is increased to $4V_o$ and the pressure is $p_o/2$, the temperature is <u>$2T_o$</u>.

8. As you drive higher in the mountains, your ears pop because the atmospheric pressure <u>decreases</u>.

9. A partial vacuum is a space of reduced <u>air pressure</u>.

10. Our bodies are dependent on pumps. Our respiratory pump is controlled by the <u>diaphragm</u> and the circulatory pump is the <u>heart</u>.

11. According to <u>Boyle's</u> law, the product of the pressure and volume of a gas at constant temperature is equal to a constant.

12. When a gas is heated in a rigid container, the <u>pressure</u> and <u>temperature</u> of the gas increases.

13. A tire pressure of "30 pounds" (per in.^{2}) is equivalent to about <u>200</u> kPa.

14. <u>High</u> pressure is generally associated with fair weather.

91

15. A liquid may be transferred from one level over an elevation to a lower level by means of a non mechanical <u>siphon</u>.

16. The inhalation and expiration of air in breathing depends on the action of the <u>diaphragm</u>.

17. A sphygmomanometer is used to measure <u>blood pressure</u>.

18. A normal blood pressure would be reported to be on the order of <u>120/80</u>.

19. Airplane lift arises because of a difference in <u>pressure (or air speeds)</u>.

20. If the air pressure in a container were 10^{-3} torr, this would be termed a <u>partial vacuum</u>.

21. According to the perfect gas law, the product of the pressure and volume is directly proportional to the <u>temperature</u> of the gas.

22. Air is composed chiefly of <u>nitrogen</u>.

23. Hot-air balloons rise because they are <u>less</u> dense than atmospheric air.

24. In a blood pressure reading of 120/80, the 120 is the systolic pressure in <u>torr</u>.

25. The lift of an airfoil or airplane wing depends on a <u>greater</u> air flow over the top surface of the foil than over the bottom surface.

26. Airplane lift is explained by <u>Bernoulli's</u> principle.

27. Altimeters are barometers with <u>inverted</u> scales.

28. Siphoning action requires a pressure <u>difference</u>.

29. As you drink through a straw, the pressure in the straw (or your mouth) is <u>less</u> than the pressure on the surface of the liquid.

30. An airplane lifts off the ground because the pressure above the wing is <u>less</u> than the pressure below the wing.

31. A helium balloon rises because helium is <u>less</u> dense than air.

32. When the temperature of a volume of gas in a rigid container is increased, the pressure of the gas <u>increases</u>.

33. A gas has no definite shape or <u>volume</u>.

34. The kinetic energy of a perfect gas is its <u>internal</u> energy.

35. The pressure times the volume of a perfect gas is proportional to the <u>average</u> kinetic energy of the gas molecules.

36. One half of the atmosphere (molecules of gas) lies below an altitude of <u>3.5</u> miles [or <u>5.6</u> km].

37. One millimeter of mercury is called a <u>torr</u>.

38. When heated air expands and increases in volume, its density is <u>lowered</u>.

39. A pump is a machine in which mechanical energy is transferred to a <u>fluid</u> causing it to flow.

40. Bernoulli's principle states that in regions where the gas speed is greater the <u>pressure</u> is less.

41. The two most abundant gases that compose the atmosphere are <u>oxygen</u> and <u>nitrogen</u>.

42. The pressure of the atmosphere <u>decreases</u> with an increase in altitude.

Chapter 10

Matching

(Choose the appropriate answer from the list on the right.)

_i_ 1. Boyle's gas law

_g_ 2. Bernoulli's principle

_m_ 3. Pascal's principle

_e_ 4. fluid

_k_ 5. perfect gas law

_h_ 6. 1 mm of Hg

_n_ 7. 1 atm

_a_ 8. barometer

_f_ 9. sphygmomanometer

_o_ 10. oxygen

_j_ 11. siphon

_c_ 12. hot-air-balloon

_l_ 13. systollic pressure

_b_ 14. nitrogen

_d_ 15. diastollic pressure

a. measures atmospheric pressure

b. 78% of atmosphere

c. functions due to bouyant force

d. minimum value

e. gas

f. used to measure blood pressure

g. relationship between fluid speed and pressure

h. 1 torr

i. applies to a constant temperature process

j. operates because of a pressure difference

k. $pV = NkT$

l. maximum value

m. explains why an open container does not collapse due to atmospheric pressure

n. 100 kPa

o. 21% of atmosphere

Chapter 11 Temperature and Heat

Answers to Questions

1. Heat is energy in transit.

2. From a substance with a higher thermal energy (per molecule), but a substance at a lower temperature may have more internal energy.

3. Heat is transferred, not cold.

4. None. The thermal energy of perfect gas is its internal energy (no molecular potential energy).

5. No. Helium is a monatomic gas and oxygen is diatomic.

6. The ice would fall off since the brass strip contracts more.

7. Thermal expansion and contraction.

8. Expansion joints to prevent cracking.

9. The temperature of the gasoline is lower than the air temperature, since the gasoline was stored in an underground tank. The hot afternoon temperature caused the temperature of the gasoline to increase which will produce expansion of the gasoline's volume and overflow will take place.

10. Stress due to thermal expansion.

11. Special materials with small thermal expansions.

12. Cold inside, hot outside.

13. Hole in ring gets larger with heating.

14. Hot water from water heater causes pipe to expand and flow rate is reduced.

15. (a) Greater displacement of molecular vibrations with heating.

16. Energy transfer to thermometer bulb by molecular collisions. Yes, the glass expands making the capillary bore larger.

17. Formation of open molecular structure.

18. A water thermometer would show a slight temperature rise due to volume expansion between 4° and 0°C. Also, not too good for measuring subzero temperatures.

19. The glass expands making the capillary bore larger.

20. Celsius.

21. (a) Fahrenheit, (b) 1 kelvin = 1.8 degrees Fahrenheit, or 1 degree Fahrenheit = 5/9 kelvin,

    (c) $1°C = 1$ K.

22. Fahrenheit.

23. 100 kelvin, $20°C + 273 = 293$ K.

24. Zero, but mass must have volume.

25. $20°C + 273 = 293$ K, then 2(293 K) = 586 K, and

    586 K − 273 = $313°C$.

26. One kelvin instead of one degree Celsius.

27. 4.2 J is the amount of (heat) energy required to raise the temperature of 1 g of water $1°C$.

28. Yes, kcal/kg = 1000 cal/1000 g = cal/g.

29. Due to grease used in cooking.

30. No, since alcohol has smaller heat of combustion.

31. Higher specific heat of apple filling (largely water).

32. Water has a high specific heat. This means the water can store large amounts of energy.

33. The container with the higher temperature change had half as much water.

34. Three times as much heat added to the copper.

35. The depth depends on heat contained or the specific heat of the material. The aluminum has the greatest specific heat and lead the smallest. From table

    13.2, the specific heat of copper is 0.093 cal/g-$°C$, so the specific heat of lead is about the same since the block depths are about equal.

SAMPLE TEST QUESTIONS

Multiple Choice

1. The human temperature sense is associated with (a) sight, (b) hearing, (c) smell, *(d) touch.

2. Heat is (a) the same as mechanical energy, (b) the same as internal energy, *(c) energy in transfer, (d) not associated with energy.

3. Thermal energy is (a) the same as internal energy, (b) heat, *(c) associated with random molecular motions, (d) potential energy.

4. The smallest temperature unit is *(a) degree Fahrenheit, (b) degree Celsius, (c) the kelvin, (d) all are the same.

5. The temperature difference between the ice and steam points of water is (a) $180^\circ C$, *(b) 100 kelvin, (c) $100^\circ F$, (d) 180 K

6. At absolute zero, the pressure of perfect gas at constant volume would be (a) one atmosphere, (b) 100 kPa, *(c) zero, (d) a negative number.

7. A cucumber has a food energy value of 5 Calories. If all this energy could be transferred to a kilogram of water, it would raise the temperature by

   (a) $5^\circ F$, *(b) 5 K, (c) $10^\circ C$, (d) $50^\circ C$.

8. Which of the following is the best fuel in terms of its intrinsic energy? (a) coal, *(b) diesel oil, (c) natural gas, (d) wood.

9. Which of the following is a unit of energy? (a) kelvin, *(b) calorie, (c) specific heat, (d) temperature.

10. A substance with a small specific heat (a) has a small heat of combustion, (b) would show a small temperature change per unit mass when a large amount of heat is added, (c) can store a relatively large amount of internal energy, *(d) would not be used for energy storage in a solar home.

11. Temperature is a relative measure of (a) transferred energy, *(b) hotness or coldness, (c) internal energy, (d) specific heat.

12. The thermal or "temperature" energy is associated with (a) rotational molecular motion, *(b) random translational molecular motion, (c) the heat of combustion, (d) vibrational molecular motion.

13. The most common means of temperature measurement is based on (a) electrical resistance, (b) radiation, (c) current in a wire junction, *(d) thermal expansion.

14. One kelvin unit is equivalent to (a) one degree Fahrenheit, (b) 1.8 degrees Celcius, *(c) 9/5 degrees Fahrenheit, (d) one Btu.

15. A "perfect" gas (a) is any normal gas, (b) is used to define the Fahrenheit scale, (c) is independent of temperature, *(d) remains a gas at any temperature.

16. When the internal energy of a perfect gas at 100 K is doubled, the Celsius would be

    *(a) $-73^{\circ}C$, (b) $100^{\circ}C$, (c) $180^{\circ}C$, (d) $273^{\circ}C$.

17. The SI unit of (heat) energy is the (a) calorie, (b) kilocalorie, *(c) joule, (d) Btu.

18. The energy released per unit mass of a substance when burned in oxygen is the (a) thermal energy, *(b) heat of combustion, (c) specific heat, (d) random molecular kinetic energy.

19. The upper limit of temperature is (a) 0 K, (b) 1000 K, (c) one million kelvin, *(d) unknown.

20. The specific heat of water is 1.0 Kcal/kg-$^{\circ}C$. If the temperature of 2.0 g of water is lowered by $10^{\circ}C$, the amount of heat removed would be (a) 10 Kcal, *(b) 20 Kcal, (c) 50 kcal, (d) 100 Kcal.

21. The amount of heat necessary to change the temperature of one gram of a substance one degree Celsius is called (a) heat capacity, (b) heat of combustion, *(c) specific heat, (d) one calorie.

22. The temperature scale used in the perfect gas law is (a) Celsius, *(b) Kelvin, (c) Fahrenheit, (d) all of these.

23. Temperature is a measure of the _____ random translational kinetic energy per molecule of a substance. (a) total, (b) minimum, (c) maximum, *(d) average.

24. Which of the following is <u>not</u> a unit of energy?
(a) calorie, (b) BTU, (c) joule, *(d) degree

25. The heat produced per unit mass of a substance when burned in oxygen is called (a) specific heat, (b) heat capacity, *(c) heat of combustion, (d) none of these.

26. Specific heat has the units of (a) calories per gram per degree Celsius, *(b) calories per gram per degree Celsius, (c) calories per gram per degree Fahrenheit, (d) joules per kilogram per degree Celsius.

27. One Celsius degree interval (a) is larger than one Kelvin degree, (b) is smaller than one Fahrenheit degree, *(c) equals one Kelvin degree, (d) none of these.

28. Absolute zero is (a) 273 ^{O}C, (b) -449.69 ^{O}F, *(c) OK, (d) none of these.

29. One BTU (a) is the amount of heat required to change the temperature of a substance one degree Fahrenheit, (b) is larger than one kilocalorie, (c) is smaller than a calorie, *(d) none of these.

30. Which of the following is not used to measure temperature? (a) thermometer, (b) pyrometer, (c) thermocouple, *(d) barometer.

Completion

1. The specific heats of substance are measured in units of <u>kcal/kg-^{O}C</u> or <u>cal/g-^{O}C</u>.

2. Alcohol and mercury thermometers cannot be used to measure extremely high temperatures because the liquids would <u>boil or evaporate</u>.

3. The principle of common thermometers is <u>thermal expansion</u>.

4. Humans emit <u>infrared</u> radiation.

5. Water boils at <u>100</u> ^{O}C or <u>212</u> ^{O}F or <u>373</u> K, and freezes at <u>0</u> ^{O}C, or <u>32</u> ^{O}F or <u>273</u> K.

6. The ice point and steam point of water are commonly called the <u>freezing</u> point and <u>boiling</u> point, respectively.

7. If the Kelvin temperature of a perfect gas is doubled, the internal energy of the gas is doubled.

8. The amount of heat needed to raise the temperature of 1 kg by 1°C is a kilocalorie.

9. A thermometer that uses the voltage difference developed across two joined wires is called a thermocouple.

10. Heat always flows from a body of higher temperature to a body of lower temperature.

11. Temperature is a measure of the average random translational kinetic energy per molecule of a substance.

12. Heat is transferred from one body to another because of a temperature difference.

13. A home thermostat generally used a bimetallic thermometer.

14. The normal human body temperature is 37 $^\circ$C.

15. The SI temperature unit is the Kelvin.

16. The intrinsic energy value of a fuel or food is expressed in terms of the heat of combustion.

17. The unit of specific heat is calories/g- $^\circ$C or kcal/kg-$^\circ$C.

18. A substance with one of the largest specific heats is water.

19. The lowest temperature is believed to be zero K (absolute zero).

20. A temperature measurement instrument that uses a comparison of visible radiation or color is called a(n) (optical) pyrometer.

21. The total energy (kinetic plus potential) contained within a body is called its internal energy.

22. Thermal expansion refers to the changes in the dimensions of substances that occur with changes in temperature.

23. One Celsius degree interval equals 1.8 Fahrenheit intervals.

24. The SI unit of all forms of energy, transferred or otherwise, is the <u>joule</u>.

25. The units for the heat of combustion are <u>cal/g or kcal/kg</u>.

26. Temperature is associated with the <u>average</u> kinetic energy of molecules.

27. Copper has a <u>smaller</u> specific heat than water.

28. A bimatallic strip bends toward the metal of <u>smaller</u> thermal expansion when heated.

29. The Celsius temperature scale has <u>100</u> degrees or divisions between the freezing and boiling points of water.

30. If the temperature of 100 degrees Celsius is doubled, the absolute temperature is then <u>473 K</u>.

31. A heated object generally <u>expands</u> because of increased molecular motion. *34-1*

32. As a nut (for a bolt) is heated from 20°C to 200°C, the size of the hole in the middle <u>increases</u>. *34-1*

decreases

modified

Chapter 11

Matching

(Choose the appropriate answer from the list on the right.)

<u>d</u> 1. Temperature

<u>h</u> 2. Heat

<u>f</u> 3. Thermal energy

<u>m</u> 4. Thermal expansion

<u>k</u> 5. Specific heat

<u>a</u> 6. Heat of combustion

<u>l</u> 7. Heat transfer

<u>c</u> 8. calorie

<u>n</u> 9. Btu

<u>j</u> 10. Kelvin

<u>b</u> 11. 273 K

<u>o</u> 12. Fahrenheit

<u>g</u> 13. $-40^{\circ}F$

<u>i</u> 14. One degree Celsius

<u>e</u> 15. Absolute zero

a. heat energy/mass

b. $0^{\circ}C$

c. the amount of heat needed to raise the temperature of 1 gram of water $1^{\circ}C$

d. a measure of the random translational kinetic energy per molecule

e. 0 K

f. random translation motions of molecules

g. $-40^{\circ}C$

h. energy in transit

i. $1.8^{\circ}F$

j. temperature scale based on absolute zero

k. cal/g-$^{\circ}C$

l. temperature difference

m. change in dimensions that occur with a change in temperature

n. the amount of heat needed to raise the temperature of 1 lb. of water $1^{\circ}F$

o. invented the mercury-lb. in-glass thermometer

Chapter 12 Heat Transfer and Change of Phase

Answers to Questions

1. Less energy transfer in molecular collisions and greater electron bonding.

2. Not really, more asthetic and value considerations.

3. The ground is a good thermal insulator with many air voids.

4. Aluminum because of greater thermal conductivity.

5. Same answer as in Question 4.

6. (a) Conductivity directly proportional to area, and (b) inversely proportional to thickness.

7. Copper increases thermal conductivity, also promotes looks.

8. Holes provide air spaces that reduce thermal conductivity.

9. It is a heat transfer process, but not in the sense that it arises from a temperature difference.

10. Air spaces between panes and storm doors and windows decreases the thermal conductivity and heat losses.

11. Mainly by reducing the flow of heat by conduction due to the air space between the main door and storm door or windows. The windows also may have double panes.

12. Without insulation convection cycles are set up between walls that promote heat transfer and loss.

13. Room heating primarily due to radiation. Draft convection takes heated air up the chimney.

14. The atmosphere and hence the Earth would heat up. (Eventually the energetic atmosphere would be lost.)

15. Yes, but the heat or warm air is maintained primarily because the greenhouse is enclosed. Ventilation is often needed on hot days and glass panes are sometimes painted white in the summer.

16. To promote heat transfer from the coils by supplying a large area for conduction to the air and also by radiation.

17. (a) Dark clothes absorb radiant energy more readily than light-colored clothes. (b) Black car. (c) Dirty (dark) snow absorbs energy, clean snow reflects.

18. Aluminum has greater thermal conductivity than metal used in older pans (e.g., iron and tin).

19. Evacuated space reduces heat transfer by conduction and convection, and silvered walls internally reflect radiation.

20. (a) To promote and prevent heat transfer by conduction and convection. (b) More blood flow and more heat to skin, but heat is more readily lost.

21. Greater temperature difference (hotter coffee), greater rate of heat loss. Adding cream right away would heat cream and prevent initial high-temperature heat loss.

22. Not really. Forced air of fan gives greater cooling by conduction and convection.

23. (a) Expanding vapors would "explode" sealed bag. (b) To protect operator from being exposed to microwaves. (c) To prevent escape of microwaves through window. (Holes of grating small compared to wavelength of microwaves.)

24. Different molecular bondings, which affect latent heats.

25. More energy (work) needed to separate molecules to form gas.

26. Less energy needed to separate molecules to form gas.

27. Evaporation from the wet bag cools water (latent heat).

28. (a) To promote evaporation and cooling (latent heat). (b) The bridge is cooled both top and bottom.

29. The wind promotes evaporation, so the wetted finger feels cool when in the wind direction (direction from which wind is blowing).

30. $-6^{\circ}F$

31. $30^{\circ}F$

32. Evaporation (sublimation) promoted by moving air, particularly in frostfree refrigerators.

33. Valley (or radiation) fogs are formed when cooling by radiation losses at night lowers the air temperature below the dew point. Sun raises temperature above dew point and fog dissipates.

34. (a) Cooler mirrow lowers moisture-ladened air to below dew point. (b) and (c) Similar to (a), i.e., air cooled below dew point. (d) No. Steam is an invisible gas. Condensed water droplets seen.

35. By evaporation of water (sublimation) from frozen coffee in a vacuum chamber.

36. Water would not evaporate as quickly. (Who wants a "wet" perfume?)

37. No, boiling is at 100°C. Greater heat input causes more vaporization and "faster" boiling.

38. Increased pressure above water lowers boiling point.

39. (a) Bubbles rise because of pressure difference (greater internal pressure) and constricted stem causes hot water to be forced upward by rising bubbles. (b) Similar to (a). Periodic eruption due to periodic filling of geyser recess with water.

40. (a) To have an operation temperature greater than 100°C. (b) An eruption of radiator liquid due to "flash" boiling.

41. Too expensive and not needed with pressurized radiator system.

42. Alcohol has a used low boiling point and it was eventually lost.

43. Snowballs formed primarily from compaction of "wet" snow. Appreciable regulation effect would require very large force (more than available from manual pressure).

44. (a) and (b) Pike's Peak. Lower boiling point and higher freezing point due to reduced pressure.

Chapter 12

SAMPLE TEST QUESTIONS

Multiple Choice

1. Gases are poor thermal conductors because (a) they
    are generally diatomic, (b) they have low specific
    heats, *(c) their molecules move around freely
    interacting only through collisions, (d) their
    average molecular kinetic energy increases with
    temperature.

2. Most home heating and cooling is done by (a)
    conduction, *(b) convection, (c) radiation.

3. Heating in a vacuum can take place by (a)
    conduction, (b) convection, *(c) radiation.

4. The energy involved in a phase change is called the
    (a) specific heat, *(b) latent heat, (c) radiant
    energy, (d) thermal energy.

5. Which of the following would you expect to be the
    best thermal conductor? (a) air, (b) water, *(c)
    aluminum, (d) concrete.

6. Latent heat is "hidden" as (a) thermal energy *(b)
    internal energy, (c) solar energy, (d) specific
    heat.

7. A solid-gas phase change is called (a) fusion, (b)
    melting, *(c) sublimation, (d) boiling.

8. The wind chill factor is (a) the air temperature,
    (b) a measure of relative humidity, (c) the
    temperature of moving air (wind), *(d) the cooling
    effect of wind on bare skin.

9. If the relative humidity is 60 percent in the
    morning and the sun warms the air, in the afternoon
    with all other things being equal, the relative
    humidity would be (a) the same, (b) higher, *(c)
    lower, (d) 100 percent.

10. To get salt water to boil at $100^{\circ}C$, you could (a)
    use a pressure cooker, (b) add more salt, (c) add
    more water, *(d) reduce the pressure.

11. Heat transfer by solids is generally by *(a)
    conduction, (b) convection, (c) radiation, (d) all
    of the preceding.

12. Monsoons occur as a result of (a) conduction cycles, *(b) convection cycles, (c) radiation cycles, (d) thermal insulation.

13. The Earth loses energy to space by means of (a) conduction, (b) convection, *(c) radiation, (d) latent heat.

14. Sunlight feels warm on the skin primarily because of (a) visible radiation, (b) microwaves, (c) ultraviolet radiation, *(d) infrared radiation.

15. The room heating from a fire in a fireplace is chiefly due to (a) conduction, (b) convection, *(c) radiation, (d) latent heat.

16. Ice, water, and steam coexist at the (a) melting point, (b) dew point, (c) boiling point, *(d) triple point.

17. The latent heat of vaporization of water is almost how many times the latent heat of fusion? (a) 3, *(b) 7, (c) 10, (d) 12.

18. If the relative humidity is 70 percent and the air temperature increases, the relative humidity *(a) decreases, (b) increases, (c) is 100 percent, (d) is 70 percent.

19. When the vapor pressure in bubbles in a liquid exceed the pressure above the liquid, then (a) the melting point is lowered, *(b) the liquid boils, (c) the temperature of the liquid increases, (d) the boiling point increases.

20. When salt is added to ice, as in making home-made ice cream, the freezing point of the mixture (a) increases with increasing pressure, (b) is equal to the dew point, *(c) is below 0°C, (d) is above the freezing point of the ice cream mixture.

21. Which of the following is not a method of heat transfer? (a) conduction, (b) convection, *(c) sublimation, (d) radiation.

22. A change of phase is a _____ process. (a) constant volume, *(b) constant temperature, (c) constant heat, (d) none of these.

23. A change of phase from a solid directly to a gas is called (a) regelation, *(b) sublimation, (c) latent heat of vaporization, (d) none of these.

Chapter 12

24. Heat transfer takes place because of a difference in (a) potential energy, (b) heat content, (c) specific heat, *(d) temperature.

25. The heat energy associated with a phase change that is involved in the work of changing the phase of the material without a change in temperature is known as (a) heat of transition, (b) specific heat, (c) heat capacity, *(d) latent heat.

26. Radiation is a method of heat transfer by means of (a) convection currents, (b) molecular interaction, *(c) electromagnetic waves, (d) all of these.

27. Heat transfer by conduction occurs chiefly in (a) plasmas, (b) liquids, (c) gases, *(d) solids.

28. The vaporization of molecules takes place (a) from liquids only, (b) from solids only, *(c) from solids and liquids, (d) at constant temperature.

29. Dew point refers to (a) pressure, (b) absolute humidity, (c) relative humidity, *(d) temperature.

30. Melting and refreezing due to applied pressure differences is called (a) sublimation, (b) regulation, (c) pressurization, *(d) regelation.

Completion

1. Metals are good conductors of electricity since they have relatively "free" electrons.

2. When fluids are heated, their volumes will generally increase and consequently their densities will decrease.

3. The two methods of heat transfer which require matter for the heat transfer are conduction and convection.

4. The radiation emitted from a body depends on its temperature.

5. Ice at 0°C is melted and the temperature of the water is then increased to 100°C. When the ice melts, its volume decreases. From 0°C to 4°C, the volume of the water decreases, and from 4°C to 100°C the volume of the water increases.

6. The energy needed to change the phase or state of a substance is called latent heat.

7. The major cooling mechanism of our bodies is evaporation or removal of latent heat.

31-8

8. The amount of heat needed to change one kilogram of water at 100°C to steam at 100°C is the latent heat of vaporization (540) kcal.

34-5

9. The effect of air motion or wind on how we feel temperature is expressed by the wind chill factor (index).

10. The relative humidity of the air increases as the temperature decreases.

31-8

11. Heat transfer by conduction in metals is due in part to "free" electrons.

12. A tile floor feels colder on bare feet than a rug because the tile has a greater thermal conductivity.

13. On-shore breezes are experienced at the beach during the day as a result of convection cycles.

14. Dark objects are good absorers (and emitters) of visible radiation.

15. Cooking in a microwave oven is due to absorption of microwave radiation by water molecules.

16. Latent heat is involved in a change of phase.

17. At the dew point temperature, the relative humidity is 100 percent.

18. With a pressure of two atmospheres, the freezing point of water is less than 0°C.

35-3

19. When table salt is dissolved in water, the boiling point of the solution is greater than that of water.

20. The "heat rays" of sunlight is infrared radiation.

21. The method of heat transfer that involves a transfer of mass is known as convection.

22. White objects re-radiate more energy than black objects.

23. The heat of fusion for a substance is generally less.

24. The transfer of heat by conduction depends on area and thickness.

25. Cooking pots and pans are made of thermal <u>conductors</u>.

26. Air is a <u>poor</u> conductor.

27. <u>Radiation</u> of heat energy can propagate through a <u>vacuum</u> as well as transparent media.

28. <u>Sublimation</u> is a change of phase directly from a gas to a solid.

29. Relative humidity has <u>no</u> units.

30. Increased pressure on a liquid <u>raises</u> the boiling point and <u>lowers</u> the freezing point.

31. Heat transfer by conduction takes place through <u>molecular</u> collisions.

32. How well a substance conducts heat depends on the electrical bonding of its <u>molecular</u> structure.

33. <u>Solids</u> are generally the best thermal conductors.

34. Nonmetal solids have relatively few free <u>electrons</u> and are poor conductors.

35. Liquids and gases are, in general, <u>poor</u> conductors.

36. The units of thermal conductivity are <u>watts/meter kelvin</u>.

37. A good absorber is also a good <u>emitter</u> of radiation.

38. The heat-retaining process of atmospheric gases (water vapor and carbon dioxide) due to selective absorption of long-wavelength terrestrial radiation is known as the <u>greenhouse effect</u>.

39. Absolute humidity has the units of <u>gram/meter^3</u> in the metric system of units.

40. The melting and refreezing due to pressure differences is called <u>regelation</u>.

Matching

(Choose the appropriate answer from the list on the right.)

__b__ 1. Heat transfer   a. calories/gram

__i__ 2. Change of phase   b. temperature difference

__d__ 3. Conduction   c. has no units

__f__ 4. Convection   d. molecular interaction

__h__ 5. Radiation   e. grams/meter^3

__m__ 6. Sublimation   f. mass transfer

__j__ 7. Regelation   g. silver

__o__ 8. Evaporation   h. electromagnetic waves

__l__ 9. Dew point   i. constant temperature process

__a__ 10. Heat of fusion

     j. melting and refreezing due to pressure difference

__e__ 11. Absolute humidity

__c__ 12. Relative humidity

     k. solid, liquid, and gas coexist

__k__ 13. Triple point

__g__ 14. Best thermal conductor

     l. relative humidity equals 100%

__n__ 15. Boiling point   m. solid to gas

     n. the temperature at which the vapor pressure of a liquid is equal to the pressure of the atmosphere

     o. slow sublimation

Answers to Questions

1. Total heat input goes into work.

2. Heat input from atmosphere to body to warm fluid; heat output from beak to cool fluid in head (latent heat to atmosphere).

3. Watts, or more likely kW for practicality.

4. No. Multicylinder engines provide smoother energy output.

5. Cylinders fire out of synchronization and combustion is less complete, thereby reducing efficiency.

6. For 6-cylinder engine, one cylinder fires every 1/3 revolution of crankshaft; for 8-cylinder engine, every 1/4 revolution.

7. To allow the up and down strokes of pistons of this 6-cylinder engine. Cam lobs for independent operation of each 12 12 valves.

8. Piston rings.

9. To warm the air and cylinders for initial combustion in cold engine.

10. At normal operation temperatures less resistance, e.g., from cold oil, etc.

11. $W = \quad Q - \quad U$, and efficiency $= W/Q_{in} =$

    $(\quad Q - \quad U)/Q_{in}$. But, $\quad Q$ = heat in - heat out, and

    in a cyclic process where the substance comes back to original conditions $\quad U = 0$.

12. Generally, the one with the higher operating temperature, or even better, the one with the greater temperature difference between the heat reservoirs. This would be the one with the greater ideal efficiency, and hopefully, greater thermal efficiency.

13. Zero in both cases. No temperature difference, no heat flow.

14. Water-cooled. Higher operating temperature. (Also, pressurized systems for this purpose.)

15. Greater than 100%.

16. The law of conservation of energy.

17. No. Energy (and first law) is conserved, heat out = heat in plus work.

18. (a) No heat engine operating in a cycle can convert heat energy completely into work. (b) No heat engine operating in a cycle can have 100% efficiency, (c) Heat will not flow spontaneously from a colder body to a hotter body. (d) The entropy of the universe increases in every natural process.

19. No. The heat input for the convection cycles comes from the energy received from the Sun, and work is done by the expanding air (and gravity).

20. First law, no. the first law is based upon the law of conservation of energy, and as long as the energy is conserved the first law is satisfied.
Second law, yes. The second law states that no heat engine can have an efficiency of 100 percent.

21. No heat pump operating in a cycle can transfer heat from a low-temperature reservoir to a high-temperature reservoir without the application of work.

22. "You can't get something for nothing" -- first law. "You can't even break even" -- second law. "I'll never sink that low" -- third law.

23. No. Heat is expelled from the refrigerator into the room -- not only that removed from the interior of the refrigerator, but also from the work input.

24. No. Energy must be expended to lower the temperature of the hot leftovers.

25. In a sense. Heat is removed through the latent heat of melting ice. However, the freezing of the ice originally required work in the removal of heat (either mechanical or natural).

26. (a) Installation costs are higher due to the large amount of metal (copper) plumbing that must be used to circulate the water. Also, the cost of an underground storage tank, if one is used. However, the temperature difference between the water and desired temperature is less, and water has a greater specific heat than air. (b) During the winter months when the air temperature is low.

27. Condensation form the air on cooling coils.

28. Ice prevents good heat exchange from air to coils. As a result, the compressor has to work harder for cooling.

29. Entropy increases, more disorder.

30. To freeze water, heat must be removed and the process is toward a more orderly state or a decrease in entropy. (However, work must be done for this to occur and the overall entropy of the universe increases.)

31. Entropy decreases since compression gives the system a greater capability to do work.

32. Yes, in the sense that the universe is gradually running down toward its heat death.

SAMPLE TEST QUESTIONS

Multiple Choice

1. When heat is added to a system, (a) the internal energy always increases, (b) work is always done by the system, *(c) the internal energy may increase and/or work may be done, (d) its temperature always increases.

2. For a heat engine, the work output (a) is greater than the heat in, (b) is always greater than the heat out, *(c) is equal to the net heat transferred to the engine, (d) does not depend on the heat out.

3. For a constant temperature (isothermal) process, the work done by or on a system of perfect gas is equal to the change in (a) internal energy, *(b) net heat transfer, (c) entropy, (d) volume.

4. The work output of a heat engine is greater than its heat out when the thermal efficiency is greater than (a) 10%, (b) 25%, (c) 35%, *(d) 50%.

5. The thermodynamic cycle for a four stroke cycle engine has how many process paths? (a) 2, (b) 4, *(c) 6, (d) 8

6. Spontaneous heat flow from a colder body to a warmer body is in violation of the (a) first law, *(b) second law, (c) third law.

7. A heat engine with 100 percent efficiency would not violate the *(a) first law, (b) second law, (c) third law.

8. The theoretical limit for the efficiency of a cyclic heat engine is given by (a) the first law, (b) entropy, (c) thermal efficiency, *(d) Carnot efficiency.

9. The thermodynamic property that gives the direction of a process is (a) internal energy, (b) thermal efficiency, (c) ideal efficiency, *(d) entropy.

10. For every natural process, the entropy of the universe, (a) decreases, (b) remains constant, (c) is destroyed in part, *(d) increases.

11. In a thermodynamic process, (a) only   Q and   U can change, (b) heat is always transferred, (c) the entropy always increases, *(d) energy is conserved.

12. The ideal efficiency of a heat engine was developed by (a) Otto, *(c) Carnot, (c) Diesel, (c) Alausius.

13. Which of the following engines involves six thermodynamic processes per cycle? (a) Diesel, *(b) four-stroke, (c) two-stroke, (d) none of these.

14. A heat engine with a heat input of 6000 J and a heat output of 2000 J would have a thermal efficiency of (a) 33%, (b) 45%, (c) 50%, *(d) 67%.

15. Which of the following is not directly involved in statements of the second law of thermodynamics? *(a) conservation of energy, (b) entropy, (c) 100 percent efficiency, (d) spontaneous heat flow.

16. Which law does not actually forbid reaching absolute zero? *(a) first law, (b) second law, (c) third law, (d) perfect gas law.

17. Entropy is a measure of (a) thermal efficiency, (b) internal energy, *(c) the capability to do work, (d) temperature.

18. The Carnot efficiency of a heat engine would be increased if (a) the thermal efficiency were increased, *(b) the temperature of the cold temperature reservoir were decreased, (c) the engine had more cylinders, (d) the entropy increased.

19. The concept of entropy was originated by (a) Otto, (b) Carnot, *(c) Clausius, (d) Newton.

20. In the "drinking bird" heat engine, the heat input comes from or the high temperature reservoir is the (a) ether, *(b) atmosphere, (c) water, (d) flock material and glass body.

21. The first law of thermodynamics is based upon the law of conservation of (a) mass, *(b) energy, (c) momentum, (d) none of these.

22. When heat is added to a closed system, the first law of thermodynamics states that the internal energy of the system may increase and/or work is (a) done on the system, *(b) done by the system, (c) done on or by the system, (d) all of these.

23. When work is done on a closed system kept at constant heat (no heat is added or removed from the system), the (a) internal energy of the system decreases, *(b) internal energy of the system increases, (c) internal energy of the system remains constant, (d) temperature of the system decreases.

24. The first law of thermodynamics states that $\Delta Q = \Delta U + W$. If the temperature of a closed system remains constant and work is done by the system then (a) ΔQ must decrease, *(b) ΔQ must increase, (c) ΔU increases, (d) ΔU decreases.

25. The second law of thermodynamics is (a) based on the first law of thermodynamics, (b) based on the law of conservation of energy, (c) based on the law of conservation of mass, *(d) none of these.

26. The second law of thermodynamics states that (a) heat cannot be converted directly into work, (b) no heat engine operating in a cycle can have an efficiency of 100%, (c) heat will not flow spontaneously from a cooler body to a hotter body *(d) all of these.

27. The third law of thermodynamics states that (a) at absolute zero molecular energy is conserved, (b) heat energy flows from a high temperature reservoir to a low temperature reservoir, (c) it is impossible to convert heat directly into mechanical energy *(d) absolute zero is impossible to obtain.

28. Which of the following is not a heat pump? (a) a refrigerator, (b) an air conditioner, (c) a heat engine in reverse, *(d) a diesel engine.

29. The first law of thermodynamics can be stated as $\Delta Q = \Delta U + W$. The symbol ΔU represents a change in (a) heat energy, (b) mechanical energy, *(c) internal energy, (d) none of these.

30. The difference between a diesel engine and a gasoline engine is (a) the type of fuel used, (b) the type of ignition, (c) cycle processes, *(d) all of these.

Completion

1. A heat engine is a device which converts heat energy to work.

2. The limit where everything in the universe would be the same temperature and there is no heat exchange is called the heat death of the universe.

3. If the volume of a thermodynamic system remains constant in a process, then $\Delta Q = \underline{\Delta U}$.

4. A measure of disorder or the availability of energy of a system is entropy.

5. The concept of an ideal heat engine was developed by Carnot.

6. The study of the general properties of heat and heat transfer is called thermodynamics.

7. The first law of thermodynamics is a statement of the conservation of energy.

8. In a diesel engine, combustion takes place as the result of compression and temperature increase.

9. A heat pump is a device that transfers heat from a low temperature reservoir to a high temperature reservoir.

10. The efficiency of a heat pump decreases as the outside temperature decreases because the air has less internal energy.

11. Heat energy added to a system goes into increasing the internal energy and/or doing work.

12. When work is done on a gas and no heat is removed from the system, the internal energy increases.

13. The work done by a heat engine is equal to the difference in the heat input and heat output.

14. An engine "stroke" refers to up or down motions of a piston.

15. A two-cycle (Diesel) engine has no intake or exhaust strokes.

16. If an engine has a thermal efficiency of 25 percent, 25 percent of the heat input goes into useful work.

17. A heat pump transfers heat from a low-temperature reservoir to a high-temperature reservoir.

18. Thermodynamically, an air conditioner is a heat pump.

19. The second law indicates the direction of a thermodynamic process.

20. Heat flows spontaneously from a hotter body to a colder body.

21. Thermodynamics deals with the transfer and actions of heat.

22. A heat pump is a device that uses mechanical energy or work to transfer heat from a low temperature source to a high temperature region.

23. A heat engine is a device that converts heat energy to work.

24. Thermal efficiency is defined as the ratio of the work out to the heat in.

25. It is impossible to obtain a temperature of absolute zero.

26. When heat is added to thermodynamic system, the internal energy may increase.

27. If all the internal energy could be removed from a gas, it would have a temperature of <u>absolute zero</u>.

28. A heat engine has a <u>smaller</u> heat output than heat input.

29. Four-stroke-cycle engines operate with additional strokes so as to <u>reduce</u> fuel waste.

30. Diesel engines do not use a spark plug for fuel <u>ignition</u>.

31. In a two-stoke-cycle eingine, <u>four</u> thermodynamic processes make up a cycle.

32. A heat engine operating in a cycle cannot convert heat energy completely into <u>work</u>.

33. The Carnot efficiency decreases as the difference in the reservoir temperatures <u>decrease</u>.

34. <u>Work</u> is required to pump heat from a cold temperature reservoir to a hot temperature reservoir.

35. The greater the entropy of a system, the <u>less</u> available energy it has.

36. When work is done by a gas, the volume of the gas <u>decreases</u>.

37. In a gas system, if the volume remains constant and the temperature decreases, then the internal energy <u>decreases</u>.

38. The efficiency of a four-cycle engine is greater than for a two-cycle engine because of <u>fuel losses</u> in the latter.

39. Temperatures used in the Carnot efficiency equation are in <u>kelvins</u>.

40. In a heat engine, the work output decreases as the heat output <u>increases</u>.

41. The ideal efficiency of a heat engine is always <u>less</u> than 100 percent.

42. Absolute zero <u>cannot</u> be theoretically obtained.

43. In every natural process, the entropy of the universe <u>increases</u>.

Chapter 13

Matching

Answers may be used one or more times.

    a.   first law of thermodynamics

    b.   second law of thermodynamics

    c.   third law of thermodynamics

__a__ 1. work = heat in - heat out

__a__ 2. $\Delta Q = \Delta U + W$

__c__ 3. It is impossible to obtain a temperature of absolute zero.

__b__ 4. Forbids perpetual motion machine.

__a__ 5. The heat added to a system with a constant volume goes into internal energy.

__b__ 6. No heat engine is 100 percent efficient.

__b__ 7. The entropy of the universe increases in every natural process.

__a__ 8. The conservation of energy applied to thermodynamics.

__c__ 9. The lower limit of temperature is OK.

__b__ 10. Heat will not spontaneously flow from a colder body to a hotter body.

Matching

(Choose the appropriate answer from the list on the right.)

h	1.	first law	*a.	thermal efficiency
o	2.	heat engine	*b.	thermodynamics
d	3.	Otto cycle	c.	process direction
l	4.	Wankel	d.	2-stroke-cycle engine
i	5.	ideal engine	e.	maximum entropy
f	6.	heat pump	f.	refrigerator
c	7.	second law	g.	disorder
g	8.	entropy	h.	conservation of energy
n	9.	third law	i.	Carnot efficiency
e	10.	heat death of universe	*j.	cold-temperature reservoir
			*k.	perfect gas
			l.	rotary engine
			*m.	high-temperature reservoir
			n.	absolute zero
			o.	room temperature

*Answers not used.

Chapter 14   Vibrations and Waves

Answers to Questions

1. Attach a spring-loaded pen to the pendulum bob so the pen will always touch a paper drawn horizontally and perpendicular to the plane of the pendulum's oscillation. Also, a light pen (e.g., a laser) could be used with light sensitive paper.

2. The maximum displacements that the pendulum makes on either side of the center point. The pendulum instantaneously stops at these points and the total energy is all potential energy.

3. (a) Amplitude increases, (b) frequency does not change since a mass oscillating on a spring has only one frequency of oscillation, (c) since the frequency doesn't change, neither does the period, $T = 1/f$.

4. (a) when in phase, the masses oscillate together. (b) When a quarter cycle is out of phase, one mass is $90°$ behind the other, e.g., when one mass is at its amplitude position, the other is at the equilibrium position or zero displacement. (c) When completely out of phase, one mass goes up and the other goes down in a half cycle.

5. The same amount of work as energy lost. If more work is done, the amplitude would increase; if less work is done, the amplitude decreases.

6. Damping is not wanted when continued motion is desired, e.g., swinging in a swing, and a pendulum on a grandfather's clock. Damping is desirable in bathroom scale dials and meter needles or indicators.

7. The springs or "shocks" absorb the energy transmitted to the auto and quickly damp the vibrations or disturbances so as to provide a smooth ride.

8. Work must be done to raise the bottle or twig.

9. EM waves carry energy in their electric and magnetic fields. The waves are generated by accelerating electrically charged particles.

10. There is greater energy transmitted to the water by the motorboat and the waves do not damp out before reaching the shore as for a rowboat.

11. If the wave speed v is constant, then if the frequency is increased, the wavelength is decreased, $f = v/\lambda$.

12. Figure (a) has greater amplitude, wavelength, and period. Figure (b) has greater frequency.

13. Longitudinal wave particles: up and down. Transverse wave particles: side to side (horizontally).

14. No, the S-waves would not pass through the liquid outer core, so only this part of the core would be detected.

15. The breaking of surf depends on the effect of the slope of the (underwater) beach on the depth of the water. The variation of the water depth with tides would affect the breaking action, e.g., for a high tide the breaking action would be closer to the beach.

16. (a) Constructive interference, the combined displacement is greater than the displacement of an individual wave. (b) Destructive interference, the combined displacement is greater than the displacement of an individual wave. (c) Total distructive interference (zero amplitude).

17. Energy cannot be destroyed (only converted to mass in special instances). The wave form is "destroyed" with the energy "stored" in the medium.

18. Waves meeting that are in phase (constructive interference) and out of phase (destructive interference).

19. Assuming equal amplitudes and frequencies, when the waves are in phase ($x = A \sin wt$ and $y = A \sin wt$), the combined wave is a straight line (at 45°). When completely out of phase ($x = A \sin wt$ and $y = -A \sin wt$), the combined wave form is a point (total destructive interference). When the waves are 90° out of phase ($x = A \cos wt$ and $y = A \sin wt$), the combined wave form is a circle. (Think of Lissajous figures.)

20. There are four loops or half-wavelengths set up in the rope for the 4th harmonic, so there are two standing wavelengths.

Chapter 14

21. Because of the free end, an antinode is at this end, so only 1/4 wavelength combinations will fit in the rod of length L, and $\lambda_1 = L/4$, $\lambda_3 = 3L/4$,

$\lambda_5 = 5L/4$, etc., so only odd harmonics are present

(1, 3, 5, etc.)

22. Because the wrench vibrates at its resonance frequency, as does the ball bat.

23. The vibration has a resonance frequency which is driven when the automobile travels at certain speeds.

24. The driving frequencies would miss energy inputs as there would not be "pushes" everytime the swing returned to its amptitude position. For example, for $2 f_o$, there would be energy input every other time.

25. The marching frequency could be the same as the resonance frequency of the bridge, and with enough energy transfer the bridge would vibrate and possibly collapse.

SAMPLE TEST QUESTIONS

Multiple Choice

1. The sine curve made by a mass oscillating on a spring or any object in simple harmonic motion is repetitious after (a) ¼ period, (b) 1/3 period, (c) ½ period, *(d) 1 period.

2. If an oscillation has a frequency of 1 Hz, it has a period of (a) ½ s, *(b) 1 s, (c) 2 s, (d) 10 s.

3. The amplitude of a wave is (a) dependent on the period, (b) the same as the wavelength, *(c) the maximum displacement of any oscillation from its equilibrium position, (d) the distance between the maximum displacement point and the minimum displacement point.

4. The magnitude of the wave velocity is equal to *(a) λ/T, (b) fT, (c) λ/f, (d) λT

5. The oscillations are parallel to the wave motion in (a) light waves, (b) water waves, *(c) sound waves.

6. Which type of wave will travel through the Earth liquid outer core? (a) transverse waves, (b) standing waves, (c) S-waves, *(d) longitudinal waves.

7. The superposition of waves that produce a combined wave form of greater amplitude than any of the individual waves is referred to as (a) harmonic motion, (b) reflection, *(c) constructive interference, (d) destructive interference.

8. The interference of two waves of equal amplitude and frequency in a string traveling in opposite directions gives rise to *(a) a standing wave, (b) a traveling wave, (c) resonance, (d) a surf.

9. If you shake a stretched rope at its second harmonic frequency, the number of wavelengths in the standing wave in the rope is (a) $\frac{1}{2}$, *(b) 1, (c) $1\frac{1}{2}$, (d) 2.

10. A stretched string is driven in resonance at its fundamental f_o. It can also be driven in resonance at a frequency of (a) $f_o/2$, (b) 2.5 f_o, *(c) $3f_o$, (d) 3.5 f_o.

11. The number of oscillations per time is given by the (a) principle of superposition, (b) antinodes, *(c) frequency, (d) wavelength.

12. When the period of oscillation decreases, the (a) amplitude increases, *(b) the frequency increases, (c) transverse waves change to longitudinal waves, (d) cause is interference.

13. Sound is *(a) a longitudinal wave, (b) an electromagnetic wave, (c) a transverse wave, (d) caused by interference.

14. Transverse waves propagate in *(a) solids, (b) liquids, (c) gases, (d) all of the preceding.

15. The magnitude of the wave velocity is given by (a) Af, (b) λ/f, (c) f/T, *(d) λf.

16. Which of the following waves will propagate through the Earth's core? (a) surface waves, (b) S-waves, *(c) P-waves, (d) transverse waves.

17. The unit of intensity is (a) J, (b) W, (c) J/m^2, *(d) W/m^2.

18. Reflection gives rise to (a) S-waves, *(b) standing waves, (c) surf, (d) electromagnetic waves.

19. Points of maximum amplitude in a standing wave are (a) out of phase, (b) overtones, *(c) antinodes, (d) locations of destructive interference.

20. When an oscillator is driven in resonance, (a) any driving frequency may be used, *(b) there is maximum energy transfer, (c) all characteristic frequencies are present, (d) the oscillation is damped.

21. The energy of a wave is (a) directly proportional to its amplitude, *(b) proportional to the square of its amplitude, (c) proportional to the square root of its amplitude, (d) none of these.

22. The principle of superposition refers to the (a) energy of a wave, (b) resonance frequency of a wave, *(c) interference of waves with one another, (d) maximum amplitude of a wave.

23. When two waves of equal amplitude and wavelength traveling in opposite directions interfere, they are continuously (a) in phase, (b) out of phase, *(c) in phase and out of phase, (d) none of these.

24. Two waves are continuously in phase if they (a) have the same frequency, (b) have the same amplitude, *(c) have the same displacement, (d) have the same intensity.

25. The speed of a wave (a) is a function of the medium, (b) is equal to the product of the frequency and wavelength, (c) is equal to the ratio of the wavelength and the period, *(d) all of these.

26. In a transverse wave the particle displacement (a) is parallel to the speed of the wave, *(b) is a right angle to the speed of the wave, (c) has no relationship with the speed of the wave, (d) is maximum at the nodes.

27. The lowest frequency of a vibrating string or air column is called (a) the first harmonic, (b) the fundamental frequency, *(c) both a and b.

28. Points of zero amplitude on a string or rope, where standing waves are present, are called (a) antinodes, *(b) nodes, (c) troughs, (d) crests.

29. Body waves from an earthquake travel through the Earth. The S-waves are _____ waves.
(a) longitudinal, *(b) transverse, (c) standing, (d) compound.

30. A/an _____ is the propagation of energy through a medium or space from a disturbance.
(a) oscillation, (b) vibration, *(c) wave, (d) all of these.

Completion

1. The frequency of a fundamental tone is 200 Hz. The frequency of the first harmonic is 200 hz and the frequency of the second harmonic is 400 hz.

2. The amplitudes of two waves are in the ratio of 2:1. The ratio of the corresponding energies is 4:1.

3. A mass oscillating under the influence of a force that obeys Hooke's law is said to be in simple harmonic motion.

4. Transverse waves give rise to shear forces.

5. The amplitude of a wave correspond to maximum particle displacement.

6. Resonance occurs when a system is driven at a natural frequency.

7. The distance between a successive crest and trough of a wave is 2 meters. The wavelength is 4 m.

8. Material waves require the following conditions: a medium and a disturbance (energy).

9. The two types of waves which result from earthquakes are P-waves which are compressional waves and S-waves which are transverse waves.

10. All waves transfer energy.

11. If an object vibrates by means of a force that obeys Hooke's law, the object is in simple harmonic motion which can be described by a sine curve.

12. When an oscillator loses energy, its motion is said to be damped.

13. If one mass oscillates upward when another moves downward, the masses are said to oscillate out of phase.

Chapter 14

14. All waves originate from <u>disturbances</u>.

15. An example of a transverse wave is <u>light, heat, wave in a rope</u>.

16. An example of a longitudinal wave is <u>sound, P-waves</u>.

17. The distance a wave travels in one period is <u>one wavelength</u>.

18. The Earth's inner core is <u>solid</u>.

19. The principle of superposition describes the effects of <u>interference</u>.

20. The zero amplitude points in a standing wave are called <u>nodes</u>.

21. A simple pendulum will experience simple harmonic motion when the <u>restoring</u> force is proportional to the <u>displacement</u>.

22. Particles at the <u>node</u> positions of a standing wave are stationary.

23. The number of oscillations per second is the <u>frequency</u> of vibration.

24. Rarefractions are regions of <u>low</u> pressure and density, whereas compressions are regions of <u>high</u> pressure and desnity.

25. <u>Longitudinal</u> waves can propagate in all phases of matter.

26. Two waves having the same amplitude and frequency interfere as a crest meets a trough. The results will be total <u>destructive</u> interference.

27. Standing waves are produced by waves traveling in <u>opposite</u> directions.

28. If the wave frequency is reduced by one-half, the period of the wave is <u>doubled</u>.

29. The diameters of the circular paths of water particles <u>decrease</u> with depth.

30. The combined amplitude of two interfering waves is greater than either amplitude in <u>constructive</u> interference.

31. If the frequency of oscillation increases, the period <u>decreases</u>.

    52 -1

32. A <u>transverse</u> wave is a shear wave.

33. Water waves are <u>circular or compound</u> waves.

34. When a crest of one wave and a trough of another wave meet, <u>destructive</u> interference occurs.

    51 - 5

35. When an oscillator vibrates in a normal mode, it is being driven in <u>resonance</u>.

Chapter 14

Matching

(Choose the appropriate answer from the list on the right.)

_g_	1.	Simple harmonic motion
_e_	2.	Amplitude
_h_	3.	Period
_a_	4.	Frequency
_k_	5.	Wave
_o_	6.	Wavelength
_j_	7.	Wave speed
_l_	8.	Transverse wave
_c_	9.	Longitudinal
_m_	10.	Intensity
_b_	11.	Principle of superposition
_d_	12.	Destructive omterferemce
_n_	13.	Standing wave
_i_	14.	Fundamental frequency
_f_	15.	Resonance

a. cycles per second

b. interference

c. wave particles oscillate parallel to the direction of wave velocity

d. superposition with a reduction in amplitude

e. maximum displacement from equilibrium position

f. maximum energy transfer

g. Hooke's law

h. time of one cycle

i. first harmonic

j. wavelength x frequency

k. propagation of energy through a medium or space

l. wave particles oscillate perpendicular to the direction of wave velocity

m. power/area

n. result of interference of two waves of equal amplitude and frequency travelling in opposite direction

o. wave speed/frequency

Chapter 15 Sound and Music

Answers to Question

1. Physically, one must have a disturbance and a medium of propagation. Depending on the definition of sound a receiver (e.g., an ear) may be considered necessary.

2. The buzzing of some flying insects comes from the disturbances of the air made by the moving wings.

3. Sound is transmitted through the glass jar. When the air is pumped from the jar, the sound is not heard as there is no medium of propagation.

4. The energy is dissipated to the surroundings.

5. Sound resonating in one of the cans is partially transmitted by the string to the other can.

6. The wind would increase the relative speed in one direction and decrease it in the other, e.g., when shouting into the wind.

7. Water molecules are lighter (H_2O: 2 + 16 = 18) than N_2 (2 x 14 = 28) or O_2 (2 x 16 = 32).

8. The speed of sound is greater in helium than in air, and with the wavelength or the vocal cavities the same, the frequency or pitch increases, $f = c/\lambda$.

9. Because of multiple reflections and reverberation time.

10. Sound is absorbed by the snow.

11. In the time it takes for the sound to travel to you, the plane has moved and is no longer at the sound-indicated position. Contrails are frozen exhaust emissions.

12. Because of reflection and refraction.

13. Yes. For a given bouncing height, there is a resonant "dribbling" or driving frequency.

14. This shows that the natural frequencies of the wall materials are in the lower part of the sound spectrum, since there are transmitted by (resonance) vibrations through the walls.

15. Yes. Waves are propagated energy through a medium or space. They possess a given frequency which when detected will be the same, higher, or lower in frequency depending on the relative motion between the source of the wave and the observer.

16. Not a great effect because the source is not moving in the direction the sound travels to the observer.

17. Yes, doubly so.

18. The light from the moving stars is Doppler shifted. In the case of the rotating Sun, the light from each side is Doppler shifted (rotation away and rotation toward).

19. The Sun has a diameter of approximately 864,000 miles. The Sun emits light waves, and its rotation may be determined by the Doppler effect, since one edge of the Sun is moving toward an observer on the Earth and the other edge is moving away from the observer.

20. No, the pressure "bow" wave trails out and downward from a supersonic plane. The boom occurs when the shock wave passes over an observer.

21. From the text, the speed of sound in water is about 4 times that in air, so v = 4 x 340 m/s = 1360 m/s.

22. For the Mach number (M), where $M = v/v_s$, subsonic:

    M < 1, but with intensity below threshold of hearing; sonic, M = 1, but audible, and supersonic: M > 1.

23. Not really. Beats arise from interference of waves at slightly different frequencies. The best of music has to do with the "time" or how fast or slow the music is played.

24. Some of the sound may be absorbed on reflection and the reflection may not be directly backwards so sound is reflected away from the observer.

25. Yes, but this would be sound with an intensity below the threshold of hearing.

26. Motorcycle, 70 - 90 dB, depending on muffler; and jack-hammer, 90 - 110 dB. (See dB table).

27. The wave speed depends on linear mass density so the wire wrapping makes the string more massive, and with v smaller and $v = f \lambda$, the string is used for low frequencies (λ constant).

28. Vibrational energy is transmitted to the table which vibrates and amplifies the sound. The bodies of string instruments use have the same effect, i.e., a sounding board.

29. Beats are pulsating fluctuations heard by musicians when their instruments are tuned to slightly different frequencies. The musicians adjust the two instruments until no beat note is heard.

30. The lighter or less dense string should not be tightened as much or have less tension (smaller wave speed and $v = \lambda f$, so f lower).

31. In an octave the frequency doubles, and if $f_2 = 2f_1$ for overtone frequencies, then the last note of an octave is the overtone of the first note of the octave.

32. The violin finger board is much smaller and the finger positions for particular notes are more easily judged.

33. Sight: shades of color. Touch: temperature (hot and cold).

34. A pure note has a single frequency. A musical note has a dominant frequency, but several overtones for quality.

35. Music has higher frequencies and these must be reproduced to give hi-fidelity (hi-quality) sound.

36. In listening to an actual musical performance, sounds coming from different sources arrive at the observer's position at slightly different times. Stereo speakers better reproduces this effect. Quadrophonic does even better, particularly with room reflections. (Speaker frequency response not considered.)

Chapter 15

SAMPLE TEST QUESTIONS

Multiple Choice

1. The upper limit of the frequency range of human hearing is (a) 20 dB, (b) 120 dB, *(c) 20 kHz, (d) 10^9 Hz.

2. The speed of sound in air on a day the air temperature is 25°C is (a) 331 m/s, *(b) 346 m/s, (c) 352 m/s, (d) 360 m/s.

3. A person hears thunder 4 seconds after seeing a lightning flash. The lightning was approximately how far away? *(a) 1.3 km, (b) 1.5 km, (c) 2.6 km, (d) 3.0 km.

4. The Doppler effect (a) is caused by resonance, (b) occurs for a person riding in a car with the car horn blowing, (c) gives rise to beats, *(d) is used in radar.

5. An interference effect involving sound is (a) reflection, *(b) beats, (c) resonance, (d) the Doppler effect.

6. The intensity for pain threshold is how many times greater than the threshold of hearing intensity? (a) 120, (b) 10^3, (c) 10^6, *(d) 10^{12}.

7. A change in the sound level intensity from 100 dB to 70 dB corresponds to a decrease in sound intensity by a factor of (a) 10, (b) 100, *(c) 1000, (d) 10,000.

8. Hearing loudness depends not only on sound intensity, but also on (a) the Doppler effect, (b) quality, (c) musical scale, *(d) frequency.

9. A tone can be distinguished from another with the same frequency and intensity by (a) beats, *(b) quality, (c) Doppler effect, (d) resonance.

10. Pianos are usually tuned *(a) to the equally tempered scale, (b) based on a standard frequency of "middle" C, (c) to the just diatonic scale, (d) with some adjacent white keys differing by tones and others by semitones.

11. A sound with a frequency of 25 Hz is in which region of the sound frequency spectrum? (a) infrasonic, (b) supersonic, (c) ultrasonic, *(d) audible.

12. The speed of sound in air is approximately (a) 330 km/s, *(b) 1/3 km/s, (c) 300,000 km/s, (d) none of the preceding.

13. Sonar range detecting is based on *(a) reflection, (b) refraction, (c) resonance, (d) beats.

14. A Doppler "blue shift" occurs when (a) the source is moving away from a stationary observer, (b) an observer is moving away from a stationary source, *(c) the observer and source are moving towards each other, (d) the observer and source are stationary.

15. The bending of sound waves is called (a) reflection, (b) resonance, (c) the Doppler effect, *(d) refraction.

16. A supersonic bow wave is formed by a jet plane flying at (a) Mach 0.5, (b) Mach 0.75, (c) Mach 0.95, *(d) Mach 1.5.

17. The threshold of hearing is *(a) 0 dB, (b) 10 dB, (c) 60 dB, (d) 120 dB.

18. An increase of 40 dB increases the sound intensity by a factor of (a) 40, (b) 400, (c) 1000, *(d) 10,000

19. The loudness of a sound depends on (a) only the intensity, (b) beats, (c) the Doppler effect, *(d) frequency and intensity.

20. The equally tempered music scale has (a) 10 notes per octave, *(b) a standard frequency reference of A = 440 Hz, (c) intervals with whole number ratios, (d) has no sharps or flats.

21. Sound waves are _____ waves. (a) transverse, *(b) longitudinal, (c) compound, (d) parallel.

22. The pitch of a sound is related directly to its (a) loudness, (b) overtones, *(c) frequency, (d) velocity.

23. A musical scale interval having a frequency ratio of 2 to 1 is a/an *(a) octave, (b) decibel, (c) harmonic, (d) equally tempered scale.

24. Sound intensity levels are measured in (a) decibels, (b) watts per square meter, *(c) both a and b.

25. The pulsating fluctuations due to the interference of waves with equal amplitudes but slightly different frequencies are called (a) overtones, *(b) beats, (c) harmonics, (d) reverberant sounds.

26. For the Doppler effect to occur there must be (a) a moving source of sound, (b) a moving listener, *(c) relative motion between a source of sound and listener, (d) all of these.

52-2

27. The speed of sound is a function of the (a) frequency of source, (b) intensity of the sound wave, *(c) temperature of the medium, (c) all of these.

52-3

28. The loudness of a sound is a function of the _____ of the sound. (a) intensity, (b) frequency, *(c) both a and b.

29. The tone quality of a musical note depends on the (a) wave form, (b) number of harmonics, (c) number of overtones, *(d) all of these.

30. A combination of sounds judged pleasing to the ear are said to be a (a) resonance, *(b) consonance, (c) dissonance, (d) none of these.

Completion

1. The audible range for sound ranges from 20 to 20,000 Hz.

51-7

2. The modulating organ which is critical in producing human voice sounds is the larynx (vocal cords).

3. The speed of sound in air at 0°C is 331 m/s, and as the temperature increases the speed of sound increases.

52-3

4. As a truck approaches a person with its horn blowing, the person hears a pitch higher than the horn frequency, and as the truck moves away from the person a lower pitch than the horn frequency is heard.

52-2

5. A speed of Mach 3 is three times the speed of sound.

6. Musicians tune their instruments by an interference effect called beats.

7. Sound intensity levels are measured in <u>dB or W/m^2</u>.

8. The three categories of musical instruments are <u>wind</u>, <u>string</u>, and <u>percussion</u>.

9. The pitch of a string on a string instrument increases with <u>tension</u>.

10. The physical analog of tone quality is <u>overtones</u>.

11. The upper limit of the sound spectrum is set by the limit of <u>material elasticity</u>.

12. The speed of sound in air increases by <u>6 m/s</u>.

    for every 10°C temperature increase.

13. Multiple reflections that give a diffuse, continuous sound produces what is called <u>reverberant sound</u>.

14. An observer moving towards a stationary sound source hears a(n) <u>increase</u> in pitch.

15. A Doppler <u>blue</u> shift indicates that an object is moving <u>towards</u> an observer.

16. When an airplane flys at a speed greater than about <u>335</u> m/s, a sonic boom can result.

17. The threshold of human hearing corresponds to <u>zero</u> dB, and sound above ≈ <u>120</u> dB can be painful.

18. Physically, the quality of sound depends on <u>wave form (overtones)</u>.

19. The standard frequency of our common musical scales is <u>440 Hz (A$_4$)</u>.

20. Moog music is generated <u>electronically</u>.

21. Sound is a <u>longitudinal</u> wave.

22. Sound requires a <u>median</u> for its propagation.

23. A sound with a frequency of 15,000 Hz is in the <u>audible</u> region of the sound spectrum.

24. <u>Ultrasonic</u> sound waves can travel long distances in water and hence are used in sonar.

25. The speed of sound in air is on the order of <u>1/3</u> km/s.

26. Reverberant sound is due to multiple <u>reflections</u>.

27. Sound <u>intensity</u> may be expressed in units of bel.

28. A decrease of 20 dB in the sound intensity level decreases the sound intensity by a factor of <u>100</u>.

29. The "do-re-me-......" of a musical scale covers a/an <u>octave</u>.

30. Ultrasound refers to sound with frequencies above <u>20 kHz</u>.

31. In general, the speed of sound is greater in materials with <u>large</u> densities.

32. The speed of light is approximately 10^6 times greater than the speed of sound in air.

33. Echoes are produced from the <u>reflection</u> of sound waves.

34. The wavelength and frequency of a sound wave are <u>inversely</u> proportional to each other.

35. An intensity level of 100 dB is <u>10^5</u> times as great as an intensity level of 50 dB.

36. When you blow across the top of an empty bottle, this is <u>closed</u> tube sound production.

37. The subjective sensory effect related to the intensity of sound is called <u>loudness</u>.

38. The subjective sensory effect related to wave frequency is called <u>pitch</u>.

39. <u>Consonance</u> is a combination of sounds judged pleasing to the ear.

40. <u>Dissonance</u> is a combination of sounds judged displeasing to the ear.

Matching

(Choose the appropriate answer from the list on the right.)

d 1. Audible frequency region

i 2. Speed of sound

f 3. Echo

n 4. Refraction

k 5. Doppler effect

b 6. Sonic boom

m 7. Beats

l 8. Loudness

a 9. Decibel

o 10. Pitch

h 11. Quality of sound

e 12. Consonance

j 13. Dissonance

c 14. Musical scale

g 15. Octave

a. sound level intensity

b. shock wave

c. defined by whole-number ratios of frequencies

d. 20 Hz to 20 kHz

e. sounds judged pleasing

f. reflected sound

g. defined by a doubling of frequency

h. related to wave form

i. function of temperature

j. sounds judged displeasing

k. used in radar

l. effect related to the intensity of sound

m. due to the interference of waves

n. bending of waves

o. related to wave frequency

Answers to Questions

1. (a) At 50-cm position, equal and opposite repul-
    sive forces on electron. A proton would have equal
    and opposite attractive forces. (b) An electron
    would be repelled back toward the 50-cm position
    and "oscillate" back and forth through this
    position. A proton would be attracted toward the
    nearer end charge.

2. Neither an electron or proton could be placed in
    equilibrium at any position due to the attractive
    and repulsive forces. An electron would move one
    way (toward the +1 C end charge) and proton would
    move in the opposite direction.

3. $e^- = 1.6 \times 10^{-19}$ C and $Q = ne^-$ or

    $n = Q/e^- = 1/1.6 \times 10^{-19} \approx 10^{-19}$ electrons

    Same number of protons since same magnitude of
    charge as electron.

4. Because there must be a separation of or a net
    charge for an appreciable or observable force.
    (Forces due to opposite charges generally cancel.)
    Effects are observed in electrostatic case as
    described in the text.

5. The forces on the charges are the same magnitude,
    but in opposite directions, e.e., the equal and
    opposite forces of Newton's third law.

6. Yes, in as much as we believe Coulomb's law holds
    everywhere.

7. The magnitude of the force would be decreased by

    one-fourth due to the $1/r^2$ dependence, $1/(2r)^2 =$

    $1/4r^2$.

8. No, the charge is still there (conserved), only
    separated.

9. Bring the object near the electroscope bulb and
    observe if the leaves diverge. If so, the object
    is charged. To determine the type of charge, place
    a known charge on the electroscope, e.g., a
    negative charge from a rubber rod. When the object
    is then brought near the electroscope, the leaves
    would diverge if the object charge is negative and
    collapse if the object charge is positive.

10. Repulsive electric force between the charged vertical fixed conductor and rotatable arm causes the arm to rotate. This type of electroscope is sturdier, not affected by air motion as with the leaf-type, and the degree of force or charge is easier to judge than with leaves.

11. Electrostatic attraction due to a separation of charge.

12. (a) The leaves would collapse as the excess charge would be conducted to ground through the finger.
(b) The leaves would collapse as excess negative charge on the electroscope would be neutralized by positive charges from the glass rod.

13. No, charge will gradually leak off.

14. Touch the electroscope bulb and bring a positively charged rod near the bulb. Negative charges would be attracted to the bulb from ground. The negative charge could be verified by bringing a charged (positive or negative) rod near the bulb and observing the leaf action.

15. The electrostatic force due to a separation of charge. On a damp or humid day, a thin film of water on an object allows for conduction and it is more difficult to charge an object than on a dry day.

16. Electrostatic forces attract the dust to the record.

17. Rub the balloon on clothing or a rug.

18. The balloon will stick to the ceiling if the attractive induced electrostatic force is greater than the weight force of the balloon.

19. (a) When an object charged by friction is brought near a door knob, an opposite charge is induced in the knob, and if the electric force is strong enough to ionize the air, a spark occurs.

20. There would be a transfer of charge to the sphere and it would then be repelled from the rod.

21. Once charge is on the plates, work must be done against the repulsive force to put more charge on the plates. With a dielectric, work is done in inducing dipoles or orienting permanent dipoles.

22. If a dielectric were conductive, a capacitor could not be charged since the dielectric would "short" the capacitor plates and a current would flow.

23. If a charged rod is brought near one of the sphere, and then the spheres separated, they would be oppositely charged, e.g., a negative rod would repel an excess of negative charge to the opposite sphere. When separated, the nearer sphere would be positively charged.

24. With a shorter conduction path, a tall tree is more likely to be struck by lightning.

25. If lightning were to strike near the pool, swimmers could be electrocuted.

26. Yes. The place is suitable for an electrical discharge to occur. For example, a very tall structure like the empire state building.

27. No, there is no lightning where there is little or no atmosphere.

28. No, Lightning rods have very sharp points where a high density electric field is provided that allows the electric charge, which is building up, to discharge slowly and the lightning bolt fails to take place.

29. The magnitude and direction of an electric field is determined by a positive test charge.

30. (a) The electric field would diverge from the area between the positive charges with no field lines at the midpoint position. The direction of the field would be away from the positive charges. (b) Same as in (a) but the direction of the field is toward the negative charges.

31. The repulsive forces cause the charges to get as far away from each other as possible, and in equilibrium they are distributed on the outside of a conductor.

32. Because the electric field is determined using a positive test charge.

33. The stronger electric field illustrated on an electric field diagram has more lines per unit area than a weaker electric field.

34. The net force inside the object due to the charges on the outside surface is zero, i.e., the forces inside cancel vectorially.

35. Gravitational shielding is not possible because there is only an attractive force in this case and the gravitational "charges" of the "shield" would influence masses inside.

36. Yes, since the charges reside on the outer surface of the sphere. It is also safe to be in contact with the outer surface of the sphere if properly insulated (see Fig. 18.21).

37. Being insulated from ground, the boy is at the same potential as the sphere and is similarly charged. As such, his hair acts like electroscope leaves.

38. Similar to a gravitational space-time field (Chapter 5) with depressions around the positions of negative charges and peaks around the positions of positive charges.

39. Electric potential energy, measured in joules, is the energy an electric charge has due to its position in an electric force field. Electric potential is the electric potential energy per charge and is measured in volts.

40. Since $V = P.E./q$, or $P.E. = qV$, and the electric potential V depends on the charge q. A charged object with twice the potential energy of another could have twice the charge and hence both objects would have the same electric potential. The electric potential of the sphere with more P.E. could be that of the other if the sphere had one-half the charge.

41. (Electric potential) $V = PE/q$, and $PE = qV$.

42. (a) coulomb   (b) volt

SAMPLE TEST QUESTIONS

Multiple Choice

1. Electric charge is measured in units of (a) volts, *(b) coulombs, (c) newtons, (d) de Graaffs.

2. If a free electron is placed midway between a positive charge and a negative charge, the electron will (a) move toward the negative charge, (b) remain stationary, (c) move at right angles to a line connecting the charges, *(d) move toward the positive charge.

41-2

3. The handle or flipper of a light switch is made of a material that is (a) conducting, *(b) insulating, (c) semiconducting.

4. A magnesium ion (Mg^{2+}) in the ionic compound $MgCl_2$

has (a) given up two protons, (b) received two protons, *(c) given up two electrons, (d) received two electrons.

5. When a hard rubber rod is rubbed with fur, the rod is negatively charged because (a) protons are transferred to the fur, (b) the rubbing motion collects charges from the air, (c) there is charging by induction, *(d) electrons are transferred to the rod from the fur.

#2-1

6. An electroscope is positively charged. If a rubber rod that has been rubbed with fur is brought near the electroscope bulb, (a) the leaves will diverge, *(b) the leaves will come together, (c) nothing happens.

42-1

7. A glass rod that has been rubbed with silk is brought near, but not touching an electroscope bulb. If the bulb is touched with a finger and the finger removed, the electroscope will then be (a) positively charged, *(b) negatively charged, (c) unchanged, (d) neutral.

42-1

8. The electric force per charge is the unit of (a) electric potential, (b) voltage, (c) electric energy, *(d) electric field.

43-3

9. The electric potential is (a) the same as potential energy, (b) force/charge, *(c) energy/charge, (d) voltage/charge.

41-4

10. Voltage refers to (a) electric force, *(b) electric potential, (c) electric induction, (d) electrical ground.

41-4

11. The electrostatic force (a) does not depend on the signs of the charges, (b) is only attractive, (c) has the unit of volt, *(d) increases when the charge separation is decreased.

41-2

144

12. The electric force between unlike charges (a) is greater in magnitude than between equal like charges, (b) is repulsive, *(c) depends on the charges on the particles, (d) follows Newton's third law.

13. The following material with the least electron mobility is (a) silver, *(b) glass, (c) mercury (d) doped silicon.

14. The leaves of an electroscope are *(a) conductors, (b) semiconductors, (c) insulators, (d) not in contact with the bulb.

15. When a negatively charged rod is brought near a suspended, neutral metal sphere, the sphere (a) is repelled, (b) shows no polarization, *(c) has a zero net charge, (d) experiences no force.

16. Lightning (a) results from static cling, (b) can only be from cloud to cloud, (c) always strikes lightning rods, *(d) gives rise to thunder.

17. An electric field (a) is expressed in units of volts, (b) points toward a positive charge, (c) is the same as electric potential, *(d) is the force/charge at points in space.

18. The direction of an electric field (a) is in the direction of the voltage, (b) is arbitrary, *(c) is away from a positive charge, (d) is perpendicular to the field lines.

19. Electrical energy is stored in *(a) an electric field, (b) a conductor, (c) electrostatic precipitator, (d) a neutrally charged body.

20. Voltage is (a) electrical force, *(b) a measure of electric potential, (c) potential energy, (d) expressed in N/C units.

21. Which of the following is not a fundamental property? (a) mass, (b) electric charge, *(c) electric force, (d) time.

22. The law of charges states that (a) unlike charges attract, (b) like charges repel, *(c) both (a) and (b).

23. Coulomb's law states (a) like charges repel and unlike charges attract, *(b) the relationship that gives the electrostatic force between two charges, (c) the direction of an electric force field, (d) the relationship between the magnitude and direction of an electric force field.

24. Electric charge is measured in (a) volts, (b) potential energy units, (c) force units, *(d) coulombs.

25. The force of attraction or repulsion between point charges is (a) proportional to the square of the distance between the charges, (b) inverse proportional to the square root of the distance between the charges, (c) inverse proportional to the magnitude of the charges, *(d) inverse proportional to the square of the distance between the charges.

26. A material in which electrons are free to move is called a/an (a) insulator, *(b) conductor, (c) semiconductor, (d) all of these.

27. The conservation of electric charge states that the net charge of a system is (a) zero, (b) greater than zero, (c) less than zero, *(d) constant.

28. Electrostatic charges can be placed on an object by (a) friction, (b) contact, (c) induction, *(d) all of these.

29. The electric field produced by an electric charge (a) has magnitude only, (b) is a scalar quantity, *(c) is a vector quantity, (d) is measured in coulombs.

30. The direction of an electric field is (a) the direction an electron would take when placed in the field, *(b) the direction a positive charge would take when placed in the field, (c) from a negative charge toward a positive charge, (d) always away from a negative charge.

31. The magnitude of the electric field is measured in units of (a) coulombs, (b) newtons, (c) newtons x coulomb, *(d) newtons/coulomb.

32. Electrical potential (a) is measured in units of force per unit charge, *(b) equals potential energy per unit charge, (c) is the same as electrical potential energy, (d) has the units of joules per newton.

146

Chapter 16

Completion

1. The magnitude of the electric force can be computed from <u>Coulomb's Law</u> and the direction can be determined from <u>law of charges</u>.

2. The centripetal force of an orbiting electron in an atom is supplied by <u>the electric force</u>.

3. Metals are good electrical <u>conductors</u>.

4. Common experienced electrostatic effects involve charging by <u>friction</u>.

5. If a positive charged object is brought near a negatively charged electroscope, the leaves of the electroscope will <u>diverge</u>.

6. Lightning rods protect buildings by conducting electrical discharges to <u>ground</u>.

7. The electric force on a charge is given by the product of the charge and the <u>electric field</u>.

8. The electric field is represented graphically by <u>lines of force</u>.

9. Particulate matter is removed from flue gases by <u>electrostatic precipitators</u>.

10. The electric potential energy of a charge is given by the product of the charge and <u>voltage or electrical potential</u>.

11. The unit of charge is the <u>coulomb (C)</u>.

12. When a rubber rod is rubbed with a piece of fur, the rod has an excess of <u>electrons (negative charge)</u> and the fur has an excess of <u>protons (positive charge)</u>.

13. The electric field is the <u>force per charge</u>.

14. A joule/coulomb is given the unit of <u>volt</u>.

15. The two most common fundamental interactions we experience are <u>gravitational</u> and <u>electrical</u>.

16. The electrostatic force between two charges is F when the distance between the charges is r. If the distance is increased to 2r, the electrostatic force is then <u>F/4</u>.

17. An electroscope is positively charged by means of a negatively charged rubber rod. This is done by the process of <u>induction</u>.

18. An atom has two more protons than electrons. The atom is called an <u>ion</u> and has a <u>positive</u> charge.

19. Lightning results from a <u>separation</u> of charge that occurs in a thunder cloud.

20. A charged balloon will stick on a wall because of electrostatic <u>induction</u>.

21. The effects produced by static and moving electric force fields (charges) are known collectively as <u>electricity</u>.

22. Electric charge is a <u>fundamental</u> property.

23. The law of charges states that like charges <u>repel</u> and unlike charges <u>attract</u>.

24. The expression $F = k\, q_1\, q_2/r^2$ is known as <u>Coulomb's law</u>.

25. Materials in which electrons are relatively free to move are known as <u>conductors</u>.

26. The fact that the net charge of a system is <u>constant</u> is called the conservation of charge.

27. The electric field arrows represent the electric force per unit <u>charge</u>.

28. The electric field has the direction that a unit <u>positive</u> charge would take if placed in the field.

29. Electric potential equals potential energy per <u>charge</u>.

30. The unit for electric potential in the SI system is the <u>volt</u>.

31. The electric force is much <u>stronger</u> than the gravitational force.

32. The constant in Coulomb's law is much <u>smaller</u> in magnitude than the universal gravitational constant in the law of gravitation.

33. When a charged object is brought near an electroscope, the charge is indicated on the electroscope by means of <u>induction</u>.

34. A capacitor with a dielectric can store <u>more</u> energy than a capacitor without the dielectric.

35. Lightning sometimes strikes twice in the <u>same</u> place.

36. The unit of electric potential in the British system is the <u>volt</u>.

37. If an ion is negatively charged, it has more electrons than <u>protons</u>.

38. Ionic solids are <u>good</u> insulators.

39. A practical application of electrostatic used to remove particulate matter from flue gases is the <u>electrostatic precipitator</u>.

40. A semiconductor has a/an <u>intermediate</u> ability to conduct an electric charge.

Matching

(Choose the appropriate answer from the list on the right.)

_d_ 1. basic electric charge

_j_ 2. law of charges

_f_ 3. Coulomb's law

_b_ 4. conductor

_k_ 5. insulator

_h_ 6. semiconductor

_m_ 7. conservation of charge

_l_ 8. induction

_a_ 9. electric field

_o_ 10. electric potential

_e_ 11. capacitor

_n_ 12. ion

_i_ 13. dielectric

_c_ 14. electrostatic precipitator

_g_ 15. joules per coulomb

a. force per unit charge

b. passes electrons freely

c. removes particulate matter

d. 1.6×10^{-9} C

e. stores electrical energy

f. $kq_1 q_2 / r^2$

g. volts

h. charge mobility is between poor and good

i. insulator between capacitor plates

j. like charges repel

k. little electron mobility

l. charging by influence of electric force field

m. the net charge is constant

n. an atom with an unequal number of electrons and protons

o. energy per unit charge

Chapter 17   Electric Current---Charges in Motion

Answers to Questions

1. Electric lights do not "burn" in the sense of combustion. This phrase probably originated when candles and oil lamps were used for lighting.

2. One object has a higher electric potential than the other and positive charge or conventional current would flow from the higher to the lower potential.

3. The electric field travels near the speed of light in a circuit and hence influences the charge throughout the circuit almost instantaneously.

4. With direct current the electrons flow is in one direction only.

5. Energy "flows" in an ac circuit. An alternating current has energy losses in the filament of a light bulb due to resistance (similar to dc current). Resistance has no directional preference.

6. Yes, in as much as the polarity of the lead wires would alternate, however, the frequency of your "ac" would be quite small.

7. Electrical energy.

8. No, electrical energy "flows" to the lamp.

9. With $q = 10$ C and $t = 2$ s, then $I = q/t = 10/2 = 5A$.

10. Emf is electric potential or voltage. The "seat" of an emf is a voltage source, e.g., a battery.

11. The battery is the source of emf and "completes the circuit through ion motion in the cell.

12. The flashlight would not work as the "backward" batteries are working against each other in the circuit.

13. No. The two battery voltages would oppose one another.

14. The chemical action of the battery is temperature dependent and is greatly reduced at low temperatures.

15. $V_{max} = V_1 + V_2 + V_3 = 1.5 + 1.5 + 9 = 12$ V when connected in series.

16. (a) The single bulb in circuit A would be brightest and (b) the bulbs in circuit C would be dimmest. The voltage across each bulb in C is one-third of that across the bulb in A.

17. The energy goes into doing work in overcoming resistance for a current to flow. This is dissipated as joule heat or perhaps as radiation as in the cases of a lamp bulb.

18. The "waste" comes from using electrical energy when not needed, e.g., leaving the lights on.

19. Rotating the knob to a lower light setting increases the resistance and less current flows, so less energy is dissipated ($I = V/R$ and $P = IV$).

20. The 75-W bulb has greater resistance. Since it is less bright, there is less current and hence more resistance ($P = IV$ and $I = V/R$). If the filaments are the same length, then the 100-W bulb filament would have a greater diameter or cross-sectional area so more current would flow (less resistance).

21. Since the 100-W bulb has a smaller resistance than the 75-W bulb (see Question 20), more current would flow through it. The voltage drop across each bulb is the same (and equal to the voltage rise of the battery).

22. In this series case, the current through both bulbs is the same. Since the 75-W bulb has a greater resistance (see Exercise 19), it has a greater voltage drop ($V = IR$).

23. Since the resistance of metal conductors generally increases with temperature, the resistance of the wires is greater, and the joule heat losses are greater (I^2R).

24. $V = 120$ V and $I = 1.4$ A is indicated on the can opener tag, hence $P = IV = (1.4)(120) = 168$ W, and the can opener uses 1.68 times as much power as a 100-W light bulb.

25. Air conditioner, water heater, stove, iron, refrigerator, and dishwasher.

26. A hot dog is electrically conductive (moisture content) and a current flows, so it is cooked aby joule (I^2R) heat. Hot dogs are wired in parallel so each have a 120 voltage drop. Also, if wired in series and a hot dog "blew out" or was poorly connected, there would be the same problem as if household circuits were in series.

27. (a) The oxygen in air would cause the hot filament to oxidize and decrease the bulb life, i.e., "burn" out more quickly. (b) Metal atoms are "boiled" off a hot filament and these atoms condense on the glass bulb giving rise to a gray spot after long periods of use. (c) Excess current causes excess joule heating and the filament melts.

28. A 120-V appliance plugged into a 240-V outlet would burn out due to the higher voltage and greater current. In the reverse situation, the appliance would probably not operate or do so inefficiently because of the low voltage and current.

29. (a) A series circuit is a switch and the load it controls. Some older Christmas tree lights are also connected in series. (b) A parallel circuit is the wiring of a family home.

30. (a) parallel circuits
(b) If homes were wired with a series circuit, when one appliance failed (open circuit) all others on that circuit would fail to operate because supply voltage is removed. Also, an extremely high supply voltage would be required to operate all appliances.

31. Auto headlights are wired in parallel, since one can blow out (or you could remove one bulb) and the other headlight still operates.

32. There are usually some wall plugs in addition to ceiling lamps. Household circuits generally handle the electrical needs for a room or two.

33. In series, the total battery voltage is
$V = V_1 + V_2 = 6 + 6 = 12$ V, and with $R = 2$ ohms,

(a) $I = V/R = 12/2 = 6$ A.
(b) In parallel, the total battery voltage is 6 V, and $I = V/R = 6/2 = 3$ A.
(c) The total resistance is $R = R_1 + R_2 = 2 + 2 = 4$ ohms, and for batteries in series, $I = V/R = 12/4 = 3$ A; and for batteries in parallel, $I = V/R = 6/4 = 1.5$ A.

34. So the circuit component is not at a high potential when the switch is off, which would be the case if the switch were in the low-voltage or ground side of the circuit.

35. Circuit breakers are more convenient and safer, since Edison base fuses can be interchanged etc.

36. The penny completes the circuit and a dangerously large amount of current could flow in the circuit, perhaps causing a fire.

37. The birds are at the same potential as the power line and there is no completed circuit (similar to the case of an insulated person touching a charged Van de Graaff generator, see Chapter 18). If a bird could touch both power lines, bye, bye birdie.

38. The sign warns about a high voltage or a high potential (for danger). There may not be a high current, but the potential is there for a dangerously high current through someone to ground who might make contact.

39. The metal parts of the car are at a high potential and you might provide a path to ground in trying to get out of the car. It is possible that the line be shorted to ground and a safety device tripped, however, the rubber tires of the car may insulate the power line from ground.

40. Not a wise or safe thing to do. It defeats the purpose of the grounding wire and its safety feature.

41. The switch is in the ground side of the circuit and the motor side of the circuit is still hot (see Question 34).

42. If the plugged-in hair dryer were dropped in a tub with you or someone in it, it would be like plugging in the tub. Radios should also be kept away from bath tubs.

43. With one hand unavailable, you would not be able to complete a circuit through the chest and give rise to possible effects as listed in Table 19.2 With only one hand, a circuit may be completed through the fingers with possible injury, but not death.

Chapter 17

SAMPLE TEST QUESTIONS

Multiple Choice

1. Electric current is measured in units of (a) volts, (b) coulombs, (c) ohms, *(d) amps.   *43-2*

2. A flow of charge in a conductor requires (a) resistance, (b) a net charge, *(c) an energy per charge difference, (d) a metal conductor.   *41.4*

3. The positive terminal of a battery is called the *(a) anode, (b) electrolyte, (c) cathode, (d) terminal voltage.

4. In a dc circuit, the electrons move (a) alternately back and forth, (b) through the circuit at a high speed, *(c) toward the battery anode, (d) the same as in an ac circuit.   *4D - 6*

5. In an ac circuit, (a) the electrons move in only one direction, (b) current flow is due to collisions of oscillating electrons, *(c) energy flows due to an alternating electric field, (d) there is no joule heating.   *42 - 6*

6. A long flashlight uses 4 D-cells in its tube. The voltage rating of the flashlight bulb is (a) 1.5 V, (b) 4 V, *(c) 6 V, (d) 8 V.

7. When a conductor in a battery circuit heats up due to joule heat, the current in the circuit (a) become alternating, (b) is unchanged, (c) increases, *(d) decreases.   *41-4*

8. The electric power used or dissipated in an appliance depends on (a) voltage, (b) current, (c) resistance, *(d) all of the preceding.   *41-4*

9. Given several commercial resistors. How would you connect them in a circuit to have the least joule heat, *(a) in series, (b) in parallel, (c) in series-parallel.

10. The electrical protection of fuses and circuit breakers depend on (a) reducing the circuit resistance, (b) increasing the circuit resistance, *(c) opening the circuit in cases of overload and shorting, (d) absorbing the overlaod energy.

11. The charge carriers in liquid conductors are (a) electrons, (b) neutral atoms, (c) protons, *(d) ions.

12. One-millionth of an ampere is a (a) milliamp, *(b) microamp, (c) kiloamp, (d) nanoamp.

13. Batteries connected in series, as opposed to being connected in parallel, in a circuit (a) would last longer, (b) have the same total voltage, *(c) supply a greater current, (d) produce the same effects.

14. The electrical resistance of metals (a) is greater for wires of larger diameters, *(b) decreases with decreasing temperature, (c) is independent of wire lengths, (d) does not affect the joule heat.

15. In a battery circuit, the electron current *(a) is toward the anode, (b) is in the same direction as the conventional current, (c) increases when batteries are connected in parallel, (d) is alternating current.

16. The unit of electric power is (a) ft-lb, (b) amp, *(c) J/s, (d) volt.

17. If the resistance in a battery circuit is reduced by one-half, the joule heat would (a) remain the same, (b) decrease by one-half, *(c) double, (d) increase by a factor of 4.

18. For resistances of different values connected in series in a circuit, *(a) the joule heat is greatest, (b) the voltage drops across each resistance is the same, (c) the currents are different through each resistance, (d) the total resistance is a minimum.

19. The lamps of modern Christmas-trees are wired (a) in parallel, (b) in series-parallel, (c) directly in series, *(d) in parallel with a shunt resistance.

20. For electrical safety, (a) fuses are wired in parallel in a circuit, (b) shocks of 1 amp of current are not a concern, (c) a larger fuse may be placed in a circuit, *(d) none of the preceding.

21. Electrical current is (a) measured in amperes, (b) the rate of flow of electrical charge, (c) the ratio of charge to time, *(d) all of these.

22. A sustained electric current requires (a) a complete circuit, (b) a voltage source, *(c) both (a) and (b).

23. Voltage is (a) not a force unit, (b) energy per charge, (c) measured in volts, *(d) all of these.

24. A direct current (dc) (a) flows in one direction, (b) may fluctuate in magnitude, (c) has electron flow toward the positive voltage source, *(d) all of these.

25. A battery is a device that converts (a) electrical energy into mechanical energy, (b) chemical energy into mechanical energy, *(c) chemical energy into electrical energy, (d) none of these.

26. The opposition of a material to electric current flow is called (a) reactance, (b) counter emf, *(c) resistance, (d) friction.

27. Ohm's law states that the current is an electrical circuit is (a) proportional to the voltage, (b) inversely proportional to the resistance, *(c) both (a) and (b).

28. Electrical power is equal to the (a) current times the voltage, (b) current squared times the resistance, (c) voltage squared divided by the resistance, *(d) all of these.

29. In a series electrical circuit, the _____ is necessarily the same for each component in the circuit. (a) voltage, *(b) current, (c) resistance, (d) power.

30. In a parallel electrical circuit the _____ is necessarily the same for each component in the circuit. (a) current, (b) resistance, (c) power, *(d) voltage.

31. In a series electrical circuit the total resistance is the _____ of the individual resistances. (a) product, *(b) sum, (c) reciprocal, (d) square.

32. Electrical power is measured in units of (a) joules, (b) joules times seconds, *(c) joules per second, (d) none of these.

Completion

1. A coulomb/second is given the name of <u>ampere</u>.

2. By Ohm's law, the ohm unit is equivalent to <u>volt/amp</u>.

3. The unit of power is the <u>watt</u>.

4. Alternate equations for electrical power in addition to P = IV are $\underline{I^2R}$ and $\underline{V^2/R}$.

5. In a parallel circuit, the <u>voltage</u> is the same across all resistances, and in a series circuit the <u>current</u> is the same through all the resistances.

6. Common household voltage is <u>110-120</u> VAC and the voltage frequency is <u>60 Hz</u>.

7. The resistance of a wire depends on the dimension of <u>area</u> and <u>length</u>.

8. The expression V = IR is called <u>Ohm's law</u>.

9. A short metal strip used in an electrical circuit to protect against overloading is a <u>fuse</u>.

10. Given three resistors. To obtain maximum resistance, they should be connected in <u>series</u>, and to obtain minimum resistance they should be connected in <u>parallel</u>.

11. Current flow requires a <u>voltage source</u>.

12. Common household current is <u>alternating</u> current.

13. The unit of current is the <u>ampere</u>.

14. A battery converts <u>chemical</u> energy to <u>electrical</u> energy.

15. Ohm's law in equation form is <u>I = V/R</u>.

16. Conventional current is in the opposite direction to <u>electron</u> flow.

17. Electrical heating effects are called <u>Joule heat</u> or I^2R.

18. In a parallel circuit, the current from a battery <u>divides</u> at the junction of several resistances.

19. Household appliances in a circuit are wired in <u>parallel</u>.

20. The third prong of a three-prong plug when plugged in a receptacle is connected to <u>ground</u>.

21. A <u>parallel</u> circuit has different paths for the current to travel. *41-7*

22. Electrons flow from a <u>high</u> potential to a <u>low</u> potential in a circuit.

23. When batteries are connected with all positive terminals together and all negative terminals together, they are connected in <u>parallel</u>.

24. As temperature of a wire conductor increases, its resistance <u>increases</u>. *41-4*

25. Three 1.5 volt dry cells are connected in series. The total voltage of the combination is <u>4.5</u> volts.

26. If one light burns out in a <u>parallel</u> circuit, all other lights will continue to burn. *41-7*

27. A five ohm resistor and a 10 ohm resistor are connected in parallel. When connected to a voltage source, a greater current passes through the <u>5</u> ohm resistor.

28. In common metal conductors, the electric current is due to "free" <u>electrons</u>.

29. A sustained electric current requires a completed circuit and a <u>voltage source</u>.

30. A battery supplies a <u>direct</u> current. *41-6*

31. The electrolyte in common automobile batteries is <u>sulfuric acid</u>.

32. The British system unit of electric power is the <u>watt</u>.

33. The voltage drops across resistors of different values connected in series are <u>unequal</u> or <u>different</u>.

34. Resistances connected in parallel have a <u>smaller</u> total resistance than when connected in series. *41-7*

35. A circuit breaker is connected in <u>series</u> in a circuit.

36. Ohm's law states that the rate of flow of charge in an electrical circuit is proportional to the <u>voltage</u>. *42-2*

37. Ohm's law states that the rate of flow of charge in an electrical circuit is inversely proportional to the <u>resistance</u>.

38. For a given voltage, a 100-W light bulb has a <u>smaller</u> resistance than a 60-W bulb.

Chapter 17

Matching

(Choose the appropriate answer from the list on the right.)

<u>f</u> 1. electric current

<u>h</u> 2. voltage source

<u>g</u> 3. direct current

<u>j</u> 4. alternating current

<u>b</u> 5. battery

<u>l</u> 6. resistance

<u>d</u> 7. Ohm's law

<u>p</u> 8. power

<u>n</u> 9. series circuit

<u>o</u> 10. safety device

<u>k</u> 11. anode

<u>e</u> 12. cathode

<u>a</u> 13. superconductor

<u>m</u> 14. drift velocity

<u>i</u> 15. joule heat

<u>c</u> 16. parallel circuit

a. zero resistance

b. converts chemical energy to electrical energy

c. voltage the same on all components

d. $I = V/R$

e. negative terminal

f. charge/time

g. current in a battery circuit

h. generator

i. $I^2 R$ losses

j. current reverses direction

k. positive terminal

l. measured in ohms

m. net electron motion

n. current the same in all components

o. circuit breaker

p. current times voltage

Chapter 18   Magnetism

Answers to Questions

1. Located at the north geographic pole, a compass needle would point toward the nearby south magnetic pole. Near the south magnetic pole, the compass would be erratic, since the magnetic field is chiefly into the ground.

2. Nails are attracted to either pole of a magnet because of the oppositely induced poles of the nails.

3. All the nails would have a north (N) pole at the point end and a south (S) pole at the head end when suspended from the south pole end of the magnet.

4. The compass reference is magnetic north and maps are referenced to "true" or geographic north. Since these are not located at the same position, corrections need to be made to give a "true" heading.

5. Compasses would point south and ion concentrations in the Van Allen Belts would be reversed, i.e., charged particles would go in the opposite directions when entering a reversed magnetosphere.

6. The iron filings become induced magnets and line up with the field.

7. By the right-hand rule, (a) east, (b) west, and (c) magnetic north, since the induced magnetic field is up and down at the side points.   With an ac current, the (circular) magnetic field around the wire reverses directions periodically so cases (a) and (b) would alternately be east-west.

8. The magnetic field of the magnet causes the magnet domains in the needle to become aligned. The stroking should be in one direction so as to promote one direction alignment.

9. The aligned magnetic domains of a magnet may be mechanically or thermally unaligned.

10. Break the bar magnet into pieces and show there is not a "separation" of poles as in the case of a separation of charges in electrostatic induction.

11. When a magnet is broken into pieces, the magnetic domains in each piece are still aligned, so each piece is a small magnet.

11. When a magnet is broken into pieces, the magnetic domains in each piece are still aligned, so each piece is a small magnet.

12. Magnetic poles are areas of concentrated magnetic fields and hence have no exact point or pole location. Think of the magnetic field of a circular current-carrying wire loop, which has a magnetic field similar to that of a bar magnet.

13. Not really. An electrically charged needle would line up with electric fields, but the Earth has no such permanent directional field.

14. (1) The needle would be attracted to the magnet. (2) When suspended from the string, the bar magnet would act as a compass and point north. (3) The magnet would support or attract the unmagnetized bar when a pole end of the magnet is placed at the unmagnetized bar's midpoint, but the unmagnetized bar would not support or attract the magnet in a similar situation.

15. It would act as a compass and the solenoid axis would line up in a north-south direction.

16. Yes. The efficiency of the electromagnet would be increased because the iron core increases the strength of the magnetic field.

17. Yes, if ac were used the magnetic field would periodically turn on and off. (Also, there would be self-induction problems.)

18. The force would still be upward since the action on the magnet reverses the direction of the B-field and the current is reversed.

19. There would be no interaction between the stationary charged ball, and the magnet, but if the ball were swinging, the moving charge would set up a magnetic field which would interact with that of the magnet.

20. Yes, moving positively and negatively charged particles would be deflected in opposite directions when traveling through a magnetic field (not parallel to the field).

21. By the right-hand fule, (a) p^+ deflected to the right and e^- to the left, (b) No effect. (c) p^+ deflected upward and e^- downward.

22. An alternating current is needed in a motor so as to provide continuous rotation of the armature or shaft. Current reversal reverses the force on the armature coils so as to avoid stable equilibrium. This is accomplished in a dc motor by a split-ring commutator (see Fig. 20.14).

23. A relay is an electromagnetic switch. A solenoid is an electromagnetic device that converts electrical energy into linear motion (kinetic energy of the core).

24. Electromagnetic devices involve electric and magnetic energy, e.g., a relay. Electromechanical devices involve electrical (and magnetic) energy and mechanical energy, e.g., a solenoid.

25. When the push button is depressed, a current flows in the circuit and activates the electromagnet, which attracts the armature and causes the bell to "ding." However, this action breaks the contact at the contact screw and opens the circuit, causing the electromagnet to be deactivated and the armature returns to the contact screw. The process repeats as long as the button is depressed and the doorbell rings.

26. When the pushbutton is depressed, the solenoid is activated and the core is set in motion. The core strikes the right tone bar giving a chime sound. Part of the solenoid energy goes into compressing the spring, and the compressed spring causes the core to return, striking the other tone bar. This "ding-dong" chime is repeated as long as the pushbutton is depressed.

27. Since an ammeter is connected "in line" or in series in a circuit, a low resistance is necessary so the ammeter does not appreciably affect the current in the circuit. Similarly, since a voltmeter is connected in parallel, it is necessary that it be a high-resistance instrument so appreciable current does not flow from the circuit through the voltmeter.

28. Because the charged particles are deflected by and "trapped" in the Earth's magnetic field.

Chapter 18

29. Because the radiowave-absorbing D and E layers present during the day disappears or weakens after the Sun goes down.

SAMPLE TEST QUESTIONS

Multiple Choice

1. A single magnetic pole (a) may be obtained by breaking a magnet in two, (b) is called a magnetic dipole, (c) would always be a north pole, *(d) has never been observed.

2. The direction of the magnetic field of a bar magnet in the vicinity of its poles is (a) always toward the south pole end of the magnet, *(b) in the direction the north pole of a compass would point, (c) always away from the north pole end of the magnet, (d) determined by a positive electric charge.

3. The SI unit of magnetic field is the (a) watt, (b) newton, *(c) tesla, (d) newton/coulomb.

4. The Earth's magnetic field (a) arises from a huge ferromagnet rock inside the Earth, *(b) is believed to be associated with currents within the Earth's core, (c) has its north pole near geographic north, (d) has always pointed in the same direction.

5. A magnetic field is produced (a) only by a permanent magnet, *(b) electric charges in motion, (c) electric charges at rest, (d) electric charges both at rest and in motion.

6. Which of the following materials is "magnetic?" *(a) nickel, (b) aluminum, (c) copper, (d) plastic.

7. The strength of an electromagnet <u>does not</u> depend on (a) the number of windings in the coil, *(b) the direction of the current in the coil, (c) the amount of current in the coil, (d) the type of material used in the core.

8. A static electric charge in a magnetic field *(a) does not experience a force, (b) produces another magnetic field, (c) is attracted toward the magnetic field lines, (d) will be set into motion by the magnetic field.

9. A motor (a) can operate only on direct current, (b) needs only an electric field to operate, *(c) converts electrical energy into mechanical energy, (d) is sometimes called a relay.

10. The aurorae *(a) occur in the ionosphere, (b) are visible lights from solar flares, (c) occur primarily near the equator, (d) are in the Van Allen belts.

11. A magnet (a) is a monopole, *(b) is a dipole, (c) always is made of Iodestone, (d) attracts aluminum.

12. At the south pole end of a bar magnet, the direction of the magnetic field is *(a) toward the south pole, (b) parallel to the end of the bar, (c) away from the south pole.

13. The magnetic field near a straight, current-carrying wire (a) is toward the wire, (b) is away from the wire, (c) is like that of a bar magnet, *(d) may be reversed by reversing the current.

14. The magnetism of a permanent magnet may be destroyed by (a) spinning, (b) the Earth's magnetic field, *(c) heating, (d) an electric field.

15. Which of the following is not a ferromagnetic material? (a) iron, (b) Alinco, (c) cobalt, *(d) chromium.

16. Electromagnetism was discovered by (a) Van Allen, *(b) Oersted, (c) Newton, (d) Galileo.

17. The Earth's north magnetic pole (a) has always been in the same location, (b) coincides exactly with the Earth's geographic north pole, (c) is near the Earth's geographic north pole, *(d) is located in Antartica.

18. An economical means to provide low voltage control of a high voltage circuit is by a (a) contact switch, (b) solenoid, *(c) relay, (d) galvanometer.

19. A simple dc motor *(a) has a split-ring commutator, (b) contains a galvanometer, (c) operates like a solenoid, (d) has a constant current direction in its armature coils.

20. Radio waves are reflected at night chiefly by the (a) D layer, *(b) F layer, (c) E layer, (d) Van Allen belts.

Chapter 18

21. Magnetism is the phenomenon produced by (a) moving electric charges, (b) an electric current, *(c) both (a) and (b), (d) neither (a) or (b).

22. Magnetic forces are (a) generated when electric currents interact, (b) vector quantities, (c) generated when moving charges interact, *(d) all of these.

23. The law of magnet poles states that (a) like poles attract, unlike poles repel, (b) like poles repel, (c) unlike poles attract, *(d) both (b) and (c).

24. The magnetic field is (a) the magnetic force per pole, (b) measured in teslas, (c) a vector quantity, *(d) all of these.

25. Oersted discovered *(a) that an electric current produces a magnetic field, (b) that magnetic poles repel and attract, (c) magnetic domains, (d) ferromagnetic materials.

26. The magnetic properties of an element, such as iron, nickel, and cobalt, depend chiefly on the magnetic fields of its (a) protons, *(b) electrons, (c) neutrons, (d) all of these.

27. An electromagnetic is a coil of wire with a/an (a) solenoid, (b) carbon core, *(c) iron core, (d) low resistance.

28. An electric motor converts (a) mechanical energy into electrical energy, *(b) electrical energy into mechanical energy, (c) alternating current into direct current, (d) direct current into alternating current.

29. The Earth's magnetosphere is (a) a symmetric region surrounding the Earth, (b) stronger at 0° latitude than at 90° latitude, *(c) affected by the solar wind, (d) all of these.

30. The ionsphere, a region of the atmosphere (a) is composed of charged particles, (b) is arranged in distinct layers, (c) varies in ion density, *(d) all of these.

Completion

1. The magnetic force on a charged particle depends on the <u>magnitude</u> and <u>velocity</u> of the charge.

2. A compass is placed near one end of a bar magnet. If the north pole of the compass points away from the magnet, then it is near the <u>north</u> pole end of the magnet.

3. The types of fields surrounding a moving charge are an <u>electric</u> field and a <u>magnetic</u> field.

4. The law of poles states that like magnetic poles <u>repel</u> and unlike magnetic poles <u>attract</u>.

5. The SI unit for magnetic field is the <u>tesla (T)</u>.

6. The region of the atmosphere in which there is a concentration of charged particles is called the <u>ionosphere</u>.

7. A magnetic field is produced by <u>moving charged</u> particles.

8. The most common ferromagnetic material used for magnets is <u>iron</u>.

9. The basic mechanism of dial ammeters and voltmeters is the <u>galvanometer.</u>

10. Donut-shaped belts of charged particles trapped in the Earth's magnetosphere are called <u>Van Allen (radiation)</u> belts.

11. Unlike magnetic poles <u>attract</u>.

12. Along with iron and cobalt, <u>nickel</u> is a major ferromagnetic material.

13. In the right-hand rule for determining the magnetic field direction, the thumb points in the direction of the <u>convectional current</u>.

14. If a conventional current flowed from right to left across this page, the direction of the magnetic field at the bottom of the page would be <u>out of the page</u>.

15. A ferromagnetic material becomes magnetic when its <u>magnetic domains</u> become aligned.

16. The limiting strength of an electromagnet depends on <u>current of I²R losses</u>.

17. A solenoid used in automobile starting systems engages the starting motor by means of a movable <u>core</u>.

18. For continuous armature rotations, a dc motor must have a <u>split-ring commutator</u>.

19. So the galvanometer coil is not burned out in an ammeter, a shunt resistor is connected in <u>parallel</u>.

20. The Earth's magnetic field is slightly deformed by <u>solar winds</u>.

21. The magnetic field is the magnetic force per <u>pole</u>.

22. Because of electric <u>spin</u>, each electron acts as a tiny magnet.

23. A solenoid with an iron core is the principle of an <u>electromagnet.</u>

24. A moving electric charge or current in a magnetic field experiences a/an <u>force</u>.

25. The magnetic north pole of the Earth is located in the <u>southern</u> hemisphere.

26. A horseshoe magnet has <u>two</u> poles.

27. The <u>south</u> pole of a compass points toward the Earth's magnetic north pole.

28. The force experienced by a moving electric charge in a magnetic field is always <u>perpendicular</u> to both the direction of the charge motion and the magnetic field direction.

29. An electric current <u>produces</u> a magnetic field.

30. A charged particle does not experience a force when traveling <u>parallel</u> to the field lines in a magnetic field.

31. Solar flares give rise to <u>aurorae</u>.

32. An ammeter has a low resistance connected in <u>parallel</u> with the meter's moving coil.

33. Iron, nickel, and cobolt are strong magnetic materials known as <u>ferromagnetic</u> materials.

34. The magnetic field around a straight current-carrying wire at a distance from the wire is in the form of a <u>circle</u>.

35. A magnetic field is produced by <u>moving</u> charges.

36. A piece of iron becomes magnetic when the magnetic <u>domains</u> become aligned.

37. A <u>relay</u> is an electromagnetic switch.

38. The history of the Earth's magnetic field direction may be studied by means of remanent magnetism in <u>rocks</u>.

39. A charge moving <u>perpendicular</u> to magnetic field lines experiences a maximum force.

40. A solenoid as used in an automobile starting system is a relay with a <u>movable</u> coil.

41. Ions and electrons trapped in the Earth's magnetic field are believed to give rise to the Earth's <u>aurora</u>.

Chapter 18

Matching

(Choose the appropriate answer from the list on the right.)

_e_	1.	magnetic poles
_i_	2.	law of poles
_f_	3.	magnetic field
_n_	4.	magnetic domain
_k_	5.	ferromagnetic material
_s_	6.	magnetic force
_d_	7.	natural magnet
_q_	8.	direction of
_b_	9.	magnetic relay
_l_	10.	solenoid
_c_	11.	magnetosphere
_p_	12.	ampere
_t_	13.	ammeter
_a_	14.	voltmeter
_r_	15.	tesla
_h_	16.	electromagnet
_o_	17.	electric motor
_m_	18.	ionsphere
_j_	19.	solar wind
_g_	20.	aurora

a. high series resistance

b. electromagnetic switch

c. Van Allen belts

d. lodestone

e. regions of concentrated magnetic strength

f. produced by an electric current

g. results from solar flares

h. solenoid with an iron core

i. like poles repel

j. high velocity ionized particles

k. iron

l. moveable core

m. in layers named D, E, F

n. group of aligned atoms

o. converts electrical energy into mechanical energy

p. defined by mutual interaction of two current-carrying wires

q. right-hand rule

r. unit of magnetic field

s. moving charge in a magnetic field

t. low shunt resistance

Chapter 19 Electromagnetic Induction

Answers to Questions

1. Motion of magnet, motion of loop with stationary magnet, closing switch in parallel loop, etc., as given in text. Also, combinations thereof and changing magnetic field due to alternating current.

2. The Earth's magnetic field is relatively weak and generator would have a very low efficiency.

3. Yes. Field lines must pass through loop.

4. (a) Combined effect since the rate of change of the magnetic field through the loop is effected by both motions. No effect since the field through the loop would not change (no relative motion).

5. No, there is no change in the field lines through the loop.

6. (a) Yes, there is a change (increase) in the number of field lines through the loop. The conventional current would be toward the reader in the moving bar in the figure so as to produce a (downward) magnetic field through the loop to oppose the change. (b) Opposite effect, i.e., the induced current is in the opposite direction so as to give a field in the upward direction that would add to the external field and oppose the change (decrease) in the number of field lines through the loop.

7. Away from the reader or towards the opposite end of the bar in the figure for an electron since the conventional current would be toward the reader so as to give a downward magnetic field to oppose the change (increase) of field lines through the loop.

8. Work must be done in some manner to induce a voltage or give energy to the charges.

9. The induced currents would be in opposite directions in the loop so as to satisfy Lenz's law since the field lines are in opposite directions for the two cases. (Field lines are away from the north pole end of the magnet and toward the south pole end.)

10. Because the induced magnetic field gives rise to epelling forces on the atomic magnets (or domains) of the material, work is required by the conservation of energy. Mechanical work is converted into electrical energy in inducing the current.

11. (a) The galvanometer needle would first deflect one way and then the other as the magnet approached and receded from the loop. (b) Less because of repulsive and attractive effects. (See Exercise 10.)

12. Self induction is due to the change in the induced magnetic field due to a current in a wire or loop itself. In mutual induction, the magnetic field is from another source, e.g., a separate loop of wire (cf. Fig. 21.4).

13. The levitation, guidance, and propulsion for MagLev trains all depend on electromagnetic induction to provide the induced magnets for each function.

14. Same voltage (amplitude) output, but the rate of "alternating" or change in polarity would double. (Self induction effects neglected.)

15. In the sense that a back emf is generated in a motor.

16. (a) Alternating. (b) Save your money. The oscillation of the magnet would be damped due to attractive and repulsive forces on the magnet. See Exercise 10. Continuous voltage output requires a continuous work input. (A case of not being able to get something for nothing.)

17. (a) Increase magnetic field so the change in the number of field lines would be greater, (b) increase the area of the loop, and (c) increase number of loops.

18. It is really a "transfer" rather than a "transform." Power or energy is transferred from one winding to the other by electromagnetic induction.

19. Yes, but coupling is promoted by iron core so as to make the transformer more electromagnetically and economically efficient. (Without core more field lines would not pass through the secondary winding.)

20. (a) No, a constant dc current gives a constant magnetic field. (Initial and final "on and off" transient currents ignored.) (b) Yes. A time-varying pulsating dc current would produce a changing magnetic field. Transformer output would be pulsating dc.

21. For an ideal transformer, the output would be the same as the input in this case. This might be called a "step-equal" or a "step-across" transformer.

22. Different output voltages may be selected by connecting to different taps, e.g., for a "center-tap" transformer, one-half of the maximum output would be obtained when the transformer is center tapped ($N_2/2$ turns).

23. $V_s = (N_s/N_p)V_p$ or $N_s = (V_s/V_p)N_p =$

    $(12,000/120)N_p = 100 \ N_p$.

24. A (voltage) step-up transform with 5 times as many windings on the secondary as on the primary: $N_s/N_p = V_s/V_p = I_p/I_s = 10 \ A/2 \ A = 5$, and $N_s = 5N_p$.

25. Both, depending on the input and output designations or connections. With $N_p = 50$ and $N_s = 500$, a step-up transformer, but with $N_p = 500$ and $N_s = 50$, a step-down transformer.

26. Mutual induction, but coils do have self-induction which gives a back emf and reduces efficiency.

27. With a 12 V input (V_p) from the battery and a 12,000 V output (V_s), then $N_s = (V_s/V_p)N_p = (12,000/12)N_p$, and $N_s = 1000 \ N_p$.

28. Leakage losses would be increased and transmission efficiency reduced.

29. The I^2R losses would be increased and transmission efficiency reduced.

30. At the place where the electrical energy is to be used, such as the family residence.

31. Yes. Leakage of energy from the very high voltage transmission lines becomes greater in a moist atmosphere.

32. Large amounts of electrical energy is used throughout the United States, especially in the Washington, D.C. to Boston area and in Chicago, and Miami in the east. San Francisco, Los Angeles, and San Diego are evident in the west. Other cities may be easily identified.

33. Electromagnetic waves are the same in the sense that they are all composed of oscillating electric and magnetic fields and travel at the speed of light in vacuum. However, they differ in frequency and wavelength.

34. Lower frequency and less energetic.

35. (a) Because of increased production of pigment in the skin for protection from the Sun. (b) The amount of melanin pigment in the skin is low. Exposure time is needed to build up the melanin pigment.

36. Clouds block or absorb infrared (heat) radiation, but transmit the burning ultraviolet radiation.

37. By absorbing part of the uv radiation before reaching the skin.

38. It causes certain substances to fluoresce and be visible. (See Chapter 23.)

39. Because the "rays" or electrons from the tube cathode striking the anode produce X-rays.

40. (a) the CAT scan can give a good cross-sectional view of the patient's body. (b) A detailed picture is not always needed by the physician, and the cost of the CAT scan is much greater than normal plate X-ray.

41. "Braking" is an acceleration, a negative acceleration or deceleration. Perhaps "decelerating rays" would be a more appropriate name.

Chapter 19

SAMPLE TEST QUESTIONS

Multiple Choice

1.   When the end of a bar magnet is brought toward a
     stationary wire loop, (a) the number of field lines
     through the loop decreases, (b) Lenz's law does not
     apply, (c) a voltage is induced in the loop, but
     not current, *(d) the induced current depends on
     the resistance of the loop.

2.   A time-varying magnetic field is produced (a) by a
     stationary magnet, *(b) when a switch is closed in
     a battery circuit, (c) by a steady dc current, (d)
     by all currents.

3.   When a wire loop is rotated in a uniform magnetic
     field, *(a) the polarity of the induced voltage
     changes every half cycle, (b) there are always
     field lines through the loop, (c) the number of
     field lines through the loop always increases, (d)
     there is no voltage induced in the loop.

4.   Which of the following would not increase the
     magnitude of an induced voltage in a coil of wire
     due to a moving bar magnet? (a) an increase in the
     number of loops in the coil, *(b) wind the coil so
     the loops are larger (large diameter), (c) move the
     magnet more slowly, (d) move the coil toward the
     incoming magnet.

5.   A generator is a device that converts (a) chemical
     energy to electrical energy, (b) electrical energy
     into heat energy, (c) heat energy into mechanical
     energy, *(d) mechanical energy into electrical
     energy.

6.   A simple dc generator has (a) the same output as an
     ac generator, (b) two slip rings, (c) no armature,
     *(d) a commutator.

7.   A transformer with more windings on the primary
     than the secondary (a) has a greater secondary
     voltage, (b) has a greater power output than input,
     *(c) is a step-down transformer, (d) none of the
     preceding.

8. A particular transformer has four times as many
    secondary windings as primary windings. This
    transformer can be used to (a) step up the power by
    a factor of 4, (b) step up the voltage and current
    each by a factor of 2, (c) step up the voltage of a
    battery by a factor of 4, *(d) none of the
    preceding.

9. Electric power transmission is done at high
    voltages so as to (a) reduce the line resistance,
    *(b) reduce the joule heat losses, (c) reduce
    leakage losses, (d) prevent people from tampering
    with the lines.

10. A suntan is a protective mechanism against which
    type of radiation? (a) infrared, (b) invisible,
    *(c) ultraviolet, (d) X-rays.

11. Electromagnetic induction was first described by
    *(a) Faraday, (b) Lenz, (c) Maxwell, (c) Ohm.

12. Back emf arises from *(a) self-induction, (b)
    mutual induction, (c) ultraviolet radiation, (d)
    batteries.

13. Mechanical energy is converted to electrical energy
    in a (a) transformer, (b) turbine, *(c) generator,
    (d) motor.   42-6

14. The voltage step-up factor of a transformer is
    given by (a) N_s, (b) N_p, (c) N_p/N_s, *(d) N_s/N_p.

15. When a magnet is brought toward a conducting loop,
    (a) there is no induction, *(b) the induced current
    produces a magnetic field opposite in direction to
    the motion of the magnet, (c) the induction effect   46-2
    would be reduced if the loop were moved toward the
    magnet, (d) the magnet would experience an
    attractive force.

16. When a loop of wire is rotated in a uniform
    magnetic field, (a) there is no magnetic induction
    because of cancellation, (b) self-induction is
    absent, (c) there would be no galvanometer
    deflection, *(d) the output would be ac.

17. A coil of wire or an inductor in an ac circuit, (a)
    has no effect on the circuit, (b) gives rise to a
    large ohmic resistance, *(c) opposes the current
    flow due to a back emf, (d) steps up the voltage
    like a transformer.

18. A step-up transformer (a) steps up the current, *(b) has fewer windings on the primary than on the secondary, (c) operates on self-induction, (d) steps down the voltage.

19. Line losses in electric power transmission are reduced by *(a) stepping up the voltage, (b) using aluminum wire instead of copper wire, (c) using long insulators on power towers, (d) corona discharge.

20. An electromagnetic wave (a) is a longitudinal wave, (b) has electric and magnetic field oscillations in the same direction, *(c) propagates in a direction perpendicular to the planes of the E and B field oscillations, (d) requires a medium for propagation.

21. A voltage can be induced into a conductor by (a) a changing magnetic field, (b) moving a conductor across a magnetic field, *(c) both (a) and (b).

22. The magnitude of the voltage induced across a conductor depends on *(a) the time rate of change of the magnetic field, (b) the resistance of the conductor, (c) the type of material the conductor is made of, (d) all of these.

23. A voltage induced in a conducting loop when there is a change in the magnetic field through the loop is known as _____ law of induction. (a) Coulomb's, (b) Lenz's, (c) Oersted's, *(d) Faraday's.

24. The voltage induced into a conductor is in such a direction that the current flow will produce a magnetic field that opposes the field that induced the voltage. This is known as _____ law. (a) Faraday's, (b) Oersted's, *(c) Lenz's, (d) Coulomb's.

25. A generator is a device that converts *(a) mechanical energy into electrical energy, (b) electrical energy into mechanical energy, (c) dc voltage into ac voltage, (d) ac voltage into dc voltage.

26. An electrical transformer can (a) increase the magnitude of ac voltages, (b) increase the magnitude of ac currents, (c) decrease the magnitude of ac voltages, *(d) all of these.

Chapter 19

27. Electromagnetic waves (a) is radiation from accelerated charge particles, (b) travel with the speed of light in vacuum, (c) consist of two vector force fields, *(d) all of these.

28. Superconducting magnets employing the principles of electromagnetic induction are involved in the _____ of the experimental MagLev train. (a) levitation, (b) guidance, (c) propulsion, *(d) all of these.

29. Electric power transmission is (a) done at low currents, (b) done at high voltages, (c) the transmission of energy, *(d) all of these.

30. _____ waves are not electromagnetic waves. (a) Light, (b) Heat, *(c) Sound, (d) X-rays.

Completion

1. The law of electromagnetic induction was formulated by Faraday.

2. Lenz's law states that an induced current is in such a direction so as to oppose the change producing it.

3. The counter emf of a motor depends on the (rotational) speed of the armature.

4. When electricity leaves a power plant it passes through a step-up transformer. Before the electricity enters your home it passes through a step-down transformer.

5. A changing magnetic field produces a changing electric field and vice versa is described by Maxwell's equation.

6. A generator is a device that converts mechanical energy into electrical energy.

7. The voltage produced in a motor in the direction opposite to the polarity of the external voltage is called a back or counter emf.

8. A step-down transformer has more windings on its primary than on its secondary.

9. The losses in electrical power transmission are primarily from I^2R (joule heat) losses and leakage losses.

10. Sunburns and tans are due to <u>ultraviolet</u> radiation.

11. A voltage induced in one circuit as a result of a change in another circuit results from <u>electromagnetic induction</u>.

12. The polarity of an induced voltage or direction of an induced current is given by <u>Lenz's Law</u>.

13. An induced current is in such a direction that it <u>opposes</u> the change that produces it.

14. To some extent, the <u>emf</u> induced in a dc motor controls its rotational speed.

15. If a split-ring commutator is used in a simple generator, the output is <u>direct current</u>.

16. A step-down transformer <u>steps up</u> the current.

17. In an ideal transformer, the <u>energy</u> is conserved.

18. Electricity is transmitted at a high voltage in order to reduce <u>I^2R losses</u>.

19. The "coil" in an auto ignition system is actually a <u>transformer</u>.

20. <u>Power frequency</u> waves are at the low frequency end of the electromagnetic spectrum.

21. A voltage may be induced in a loop of wire by <u>turning</u> it in a uniform magnetic field.

22. The thrusting of a magnet into a coil of wire is an example of converting <u>mechanical</u> energy into <u>electrical</u> energy.

23. Microwaves have <u>lower</u> frequencies than visible light.

24. Electromagnetic waves require no material <u>medium</u> for propogation.

25. FM radio waves have <u>greater</u> frequencies than AM radio waves.

26. A step-up transformer steps down the <u>current</u>.

27. X-rays are produced in X-ray tubes by the <u>deceleration</u> of electrons.

28. A transformer will operate on <u>alternating</u> voltages.

180

29. A change in the magnetic field through a conducting loop induces a <u>voltage</u> in the loop.

30. The law of electromagnetic induction is known as <u>Faraday's</u> law.

31. A back emf in motor coils results from <u>self-induction</u>.

32. Electromotive force is a <u>voltage</u>.

33. An alternator is the same as a/an <u>ac</u> generator.

34. A transformer operates on the principle of <u>mutual</u> induction.

35. A <u>step-up</u> transformer has more windings on the secondary than on the primary.

36. Leakage losses in power transmission result from losses by <u>ionization</u>.

37. <u>Ultraviolet</u> radiation causes suntans and sunburns. *54-4*

38. Lenz's law is a corollary of the law of conservation of <u>energy</u>.

Matching

(Choose the appropriate answer from the list on the right.)

d 1. Faraday's law

h 2. Lenz's law

j 3. electromagnetic induction

f 4. generator

n 5. step-up transformer

p 6. electromagnetic waves

l 7. self-induction

a 8. MagLev

o 9. electric power transmission

e 10. CAT scan

b 11. line losses

i 12. ultraviolet radiation

g 13. alternator

k 14. step-down transformer

m 15. TV waves

c 16. infrared radiation

a. levitation

b. I^2R

c. heat rays

d. law of induction

e. x-ray technique

f. mechanical energy to electrical energy

g. ac generator

h. gives the direction of induced current

i. absorbed by ozone layer

j. induced voltage as a result of time-varying magnetic field

k. increases current

l. back emf

m. MHz range

n. increases voltage

o. high voltage transfer

p. radiation from accelerating electric charge

Chapter 20 Light Waves

Answers to Questions

1. In the sense that when the received light intensity is too low to stimulate a vision response and things are "dark".

2. The vibrations of a quartz crystal are used for a time standard, similar to the oscillations of a pendulum of a grandfather clock.

3. The "life times" of the excited states of the atoms of the material determines how long a material will glow. The brightness depends on the number of excited states.

4. The light would be "dimmer" because of a shift in the spectral frequency distribution toward the red end of the spectrum.

5. Mercury vapor has transitions in the blue-uv end of the spectrum. Sodium has intense transitions in the yellow region of the spectrum.

6. No, the fluorescent effects require uv light and infrared light is called "heat" rays (Chapter 7)., so you might have some hot band members as well as "hot" music.

7. "Black" lights emit some light in the blue end of the visible spectrum as well as uv radiation.

8. As the hint states, a slight blue tint to white clothes is preferred and the blue fluorescent dye gives this, when viewed under fluorescent lamps.

9. Due to dried soap films on the dish surfaces, giving rise to thin film interference.

10. The film would appear either the color of the monochromatic light or dark depending on whether the thin film interference was constructive or destructive.

11. No. It depends on the phase shifts at the reflecting surfaces. For a soap film there is a

    180° phase shift at the external air-film interface, but no phase shift at the film-internal air interface. As a result, destructive interference occurs for a film thickness of $\lambda/2$.

12. The eye is more sensitive to yellow-green light and the nonreflection of light in this region gives the maximum effect. If the film thickness were $\lambda/2$, the glass would be reflective.

13. Thinner, since blue light has a shorter wavelength.

14. Light is energy, so it can't be "destroyed." Destructive interference for thin films means that the light is transmitted. In a double slit experiment, the light that would ordinarily be in a dark fringe area is in a bright fringe area.

15. Nonreflecting lenses reduce reflection or improve transmission for better film exposure. Also, back reflections are reduced that could give rise to poor images. See Special Feature 23.1.

16. Both reflections are from a less "optically dense" medium, so there is no phase shift. With a $\lambda/4$ film thickness or path difference of $\lambda/2$, the waves interfere destructively and back reflection is reduced.

17. (a) Film more optically dense than glass. Air-film reflection is phase shifted 180° and film-glass reflection is not phase shifted, so a film thickness of $\lambda/4$ would give a path difference of $\lambda/2$ resulting in constructive interference and reflection. (b) Film less optically dense than glass. Both interface reflections phase shifted by 180°, so a film thickness of $\lambda/2$ would give a path difference resulting in constructive interference and reflection.

18. The alternate path differences of $\lambda/4$ and $\lambda/2$ of the air wedge would produce alternate lines. Incident white light would give colored lines resulting from interference for the wavelengths of the different light components and the appropriate constructive and destructive conditions.

19. The feathers act as diffraction grating.

20. Reflected light is polarized in the horizontal plane. By orienting the transmission axis of the sunglass lenses vertically, the reflected polarized light is blocked and the intensity (glare) reduced.

21. See Question 22.

22. If both are polarizing sunglasses, this could be shown by a cross polarizing condition. If one or neither were polarizing, you could not tell by the previous condition, but you might by rotating the lense and reviewing reflected light or scattered skylight to note variations in the intensity of the transmitted light.

23. (a) Twice per revolution, (b) Four times. (c) No variation.

24. No. Only one image would be seen through regular sunglasses with both lenses transmission axes vertical.

SAMPLE TEST QUESTIONS

Multiple Choice

1. An object that "glows in the dark" after the lights are shut off is an example of (a) joule heating, (b) polarization, (c) interference, *(d) phosphorescence.

2. The excited mercury vapor in a fluorescent lamp emits what type of radiation? (a) infrared, *(b) ultraviolet, (c) visible, (d) heat.

3. Young's double slit experiment *(a) allows measurement of light wavelengths, (b) is an example of fluorescence, (c) applies only to ultraviolet radiation, (d) involves polarization.

4. The film thickness of nonreflecting glass 1/4 of a wavelength (a) because only one reflection is phase shifted, *(b) so as to give a path difference for destructive interference, (c) to avoid polarization, (d) to promote diffraction.

5. Newton's rings arise due to *(a) interference, (b) diffraction, (c) polarization, (d) fluorescence.

6. The bending of waves around corners is called (a) interference, (b) reflection, *(c) diffraction, (d) polarization. 51-4

7. A doorway or open window would greatly diffract (a) light waves, *(b) sound waves, (c) AM radio waves, (d) TV waves.

8. A diffraction grating can be used to (a) polarize light, (b) produce Newton's rings, *(c) separate light into a spectrum, (d) produce light.

185

9. Polarization involves *(a) orientation of field vectors, (b) bending of light around corners, (c) interference, (d) longitudinal waves.

10. LCD's using a twisted nematic display involves (a) interference, (b) diffraction, (c) fluorescence, *(d) polarization.

11. The color of light emitted from gas discharge tubes, e.g., a neon tube, depends on (a) polarization, *(b) orbital spacings, (c) interference, (d) diffraction.

12. Certain objects glow in the dark because of (a) polarization, (b) interference effects, *(c) delayed electron transitions, (c) infrared radiation.

13. Incandescent lamps produce predominantly (a) monochromatic light, (b) polarized light, (c) visible light, *(d) infrared radiation.

14. The gas of fluorescent lamps emits *(a) ultraviolet radiation, (b) infrared radiation, (c) visible light, (d) polarized light.

15. Young's double slit experiment (a) produces polarized light, *(b) infrared radiation, (c) visible light, (d) polarized light.

16. Sound waves (a) are ordinarily monochromatic, (b) can be polarized, *(c) are defracted by normal openings, e.g., door, (d) do not show interference effects.

17. Reflection gratings (a) have only a few lines, (b) depend on internal reflection, (c) polarize light, *(d) are opaque.

18. The colors seen in soap bubbles (a) result from diffraction, (b) are in the ultraviolet region, (c) occur when the bubble thickness is uniform, *(d) involve destructive interference.

19. Polarization can result from *(a) scattering, (b) interference, (c) diffraction, (d) fluorescence.

20. When voltage is applied to a liquid crystal, (a) polarized light cannot be transmitted, (b) polarized light is rotated, *(c) the region of voltage application appears dark in a LCD, (d) the crystal properties are destroyed.

21. The process whereby certain materials absorb ultraviolet light and emit light in the lower-frequency visible region is known as (a) phosphorescence, (b) diffraction, (c) polarization, *(d) fluorescence.

22. Light waves experience (a) diffraction, (b) polarization, (c) interference, *(d) all of these.

23. Light which is most sensitive to the human eye is the _____ portion of the visible spectrum. (a) red-orange, (b) orange-yellow, *(c) yellow-green, (d) blue-green.

24. Light (a) has a dual nature, (b) is electromagnetic radiation, (c) can be polarized, *(d) all of these.

25. Light is generated when *(a) electrons are accelerated, (b) diffraction occurs, (c) polarization takes place, (d) none of these

26. Phosphorescent materials are excited by (a) diffraction, (b) polarization, *(c) exposing them to light, (d) joule heating.

27. The change in direction of a light wave due to bending of the wave around a sharp edge is called (a) refraction, *(b) diffraction, (c) polarization, (d) interference.

28. The incandescent lamp as a light source _____ a fluorescent lamp. (a) is more efficient than, (b) has about the same efficiency as, *(c) is less efficient than.

29. When the field vectors of light waves are preferential orientated _____ will occur. (a) diffraction, *(b) polarization, (c) fluorescence, (d) phosphorescence.

30. The effects of _____ are made use of in the operation of liquid crystal displays. (a) diffraction, (b) fluorence, (c) light wave interference, *(d) polarization.

Completion

1. Slit diffraction depends on the wavelength and size of the slit.

2. Three ways to polarize light are by means of selective absorption, scattering, and reflecting.

3. Fluorescent materials fluorescence when <u>ultraviolet</u> light is absorbed.

4. Interference, diffraction, and the Doppler effect are explained in terms of <u>wave</u> properties.

5. When atomic electrons "jump" from a higher energy level to a lower energy level, <u>radiation (light)</u> is given off as energy.

6. Light from a neon gas consists mainly of <u>red</u> light.

7. "Black" light consists mainly of <u>ultraviolet</u> radiation.

8. When a film thickness is such that the light reflected from the upper and lower surfaces are in phase, <u>constructive</u> interference occurs.

9. When two waves interfere that are out of phase, <u>destructive</u> interference occurs.

10. The wavelengths of sound are <u>greater or longer</u> than those of visible light.

11. Luminous toys and clock dials glow in the dark because of <u>phosphorescence</u>.

12. <u>Incandescent</u> lamps are very inefficient with about 95 percent of the energy radiated as heat rays.

13. The dark fringes in Young's double slit experiment are a result of <u>destructive interference</u>.

14. The colored regions seen in a soap bubble result from <u>constructive interference</u> for particular wavelengths.

15. Diffraction effects are greater when the object or opening size is <u>the same order or smaller</u> as the wavelength of the wave.

16. Etched, parallel lines on a piece of glass form a <u>diffraction grating</u>.

17. Atoms that absorb ultraviolet light and reemit visible light are said <u>fluoresce</u>.

18. FM radio reception may be blocked by buildings because the wavelengths are too long to be <u>diffracted</u> around the buildings.

19. The transmission axis of a Polaroid film is <u>perpendicular</u> to the orientation direction of the molecular chains.

20. Liquid crystals in LCD's without an applied voltage rotate the polarization direction of polarized

    light by <u>90</u>$^{\circ}$.

21. Electrons must <u>oscillate</u> many billions of times per second to radiate visible light.

22. Light of multiple frequencies is called <u>polychromatic</u>.

23. Fluorescence generally involves the absorption of <u>ultraviolet</u> light.

24. TV's sometimes flutter when airplanes fly over as a result of reflection and <u>interference</u>.

25. <u>Diffraction</u> is commonly described as a "bending" of waves around sharp edges. $51-4$

26. Unpolarized light has <u>randomly</u> oriented field vectors.

27. Newton's rings result from <u>interference</u>.

28. Polarizing sunglasses have the transmission axis oriented <u>vertically</u>.

29. A dark fringe in a double slit interference pattern is the result of <u>destructive</u> interference. $51-5$

30. The propagation direction of light is <u>perpendicular</u> to both the magnetic and electric field vectors.

31. Polarization applies to <u>transverse</u> waves.

32. The direction perpendicular to the oriented molecular chains in a polarizing sheet is the <u>polarization</u> direction.

33. <u>Polychromatic</u> light passing through a diffraction grating gives rise to spectra in the bright fringes.

34. The light coming from LCD's is <u>polarized</u>.

35. If monochromatic light is incident on a soap film, there will be alternating <u>light</u> and <u>dark</u> areas seen on the film.

36. <u>Energy</u> is emitted from excited atoms when they de-excite.

37. Most of the energy emitted from an incandescent lamp is <u>heat</u> radiation.

38. An incandescent lamp emits light as a result of <u>electrical</u> heating.

Matching

(Choose the appropriate answer from the list on the right.)

 n 1. Light

 d 2. Phosphorescence

 h 3. Fluorescence

 f 4. Newton's rings

 l 5. Diffraction

 j 6. Polarization

 o 7. Constructive interference

 a 8. Destructive interference

 m 9. Young's experiment

 k 10. Monochromatic

 g 11. LCD

 b 12. Nonreflecting film

 i 13. Polychromatic

 c 14. Majority of the radiation produced by incandescent lamps

 e 15. Dual nature of light

a. produces dark fringes

b. one-quarter wavelength

c. infrared

d. glows in the dark

e. waves versus particles

f. lenses irregularities

g. twisted nematic display

h. ultraviolet radiation

i. light containing two or more frequencies

j. field vector orientation

k. light of one color

l. the "bending" of light

m. wavelength determination

n. electromagnetic radiation sensitive to the eye

o. produces bright fringes

Chapter 21 Reflection and Refraction

Answers to Questions

1. Geometrical optics is the study of light-wave phenomena geometrically by light rays. Wave or physical optics is the study of light-wave phenomena as waves.

2. A wave front is defined by the adjacent points on a wave that are in phase.

3. Yes. A ray is a line drawn perpendicular to the wave front in the direction the wave is traveling.

4. A beam of light is a group of parallel rays.

5. Rays are not reflected "sideways" to the plane of incidence.

6. (a) Reflection directly back along line of incidence (or 180° reflection). (b) No reflection in that incident light is along or parallel to the surface of the mirror.

7. Brighter because of less diffuse reflection, but dark surface would still absorb light.

8. In a dark room, a laser beam is not normally seen, only the laser "spot" on the wall due to reflection. By placing dust in the beam path, the light is diffusely reflected and the beam is visible (light reflected to the eye). Dark parts of the beam in the figure have insufficient dust particles for reflections to be visible. (Note: the dusted beam can also be seen with the classroom lights on when this is done as a demonstration.)

9. Reflections from the front and back surfaces of the glass.

10. Your image approaches you at the same speed, but is out of step, i.e., steps off with left foot when you step off with the right foot.

11. One-way mirrors are partially silvered so part of the incident light is reflected and part is transmitted. As mentioned in the hint, window-panes are effectively one-way mirrors at night.

12. Let R be the reflectivity (0.95) of mirror 1. Then from mirror 2, the reflectivity is R x R = R², etc., so for four mirrors the total reflectivity is

    $R_t = R^4 = (0.95)^4 = 0.81$.

13. A regular plane mirror gives the driver a rear view, but with a diverging spherical mirror, the driver has an expanded rear view, even though the image may be smaller and distorted.

14. With the light source at the focal point, the reflected beam is parallel for uniform illumination.

15. A converging mirror gives a magnified image so the dentist can better see small objects.

16. The wavelength is shorter since the wave velocity is less, $\lambda = v/f$ (frequency f unchanged).

17. (a) No, Θ is also $0°$, and light is partially reflected and transmitted directly. (b) At an

    incident angle of $90°$, the light beam is along or parallel to the surface of the medium.

18. The light is refracted one way (toward the normal) at the water-glass boundary and the other way (away from the normal) at the glass-air boundary, making it appear (by straight line ray thought) that the submerged part of the pencil is displaced.

19. Because of water motion and temperature differences that affect the refraction. This is similar to the atmospheric effects that cause stars to "twinkle."

20. When the pool is full of water, light from the letter near the observer are refracted or bent away from the normal or toward the observers so these letters are visible.

21. Daylight hours would be shorter if there were no atmosphere, since by atmospheric refraction the Sun is seen before it rises and after it sets.

22. The refraction at the glass boundaries makes the liquid appear closer to the outer surface of the mug or that the mug is fuller or holds more.

23. No, the refraction is toward the normal in an optically denser medium and away from the surface, so light cannot be internally reflected.

24. $\theta_2 = \theta_c$ for the glass since by reverse ray tracing greater angles of refraction are not possible.

25. Yes. Internal reflection in prisms are an example. Light from an object enters the prism, is internally reflected, and an image of the object is seen in the prism when viewing the transmitted light.

26. (a) Set plane mirrors at 45° angles for a 90° reflection ($\theta_i + \theta_r = 45^\circ = 90^\circ$) down periscope tube and a 90° reflection out of the periscope tube. (b) 90° prisms at the proper angles gives the same effect.

27. (a) No, only limited cone with the half-angle $\theta_2 = \theta_c$ for water. (Consider reverse ray tracing.) (b) A 180° view would be seen when viewing through the cone defined by θ_c. The 180° view would be "compressed" into the θ_c cone, giving a "fish-eye" lens view of the above-water world. Beyond θ_c would be a mirror view of the objects below the surface of the water due to internal reflection.

28. The windowpane thickness or light path is not great enough for the colors to be dispersed appreciably.

29. Another single prism would not recombine the light components, but also disperse them. [Three more prisms are actually needed to converge the dispersed light from the original prism. Essentially, prism combinations act as converging a lens.]

To synthesize the dispersed light four prisms are required. Arrange the first two prisms with parallel bases vertically and have the white light beam horizontally incident slightly above or below the apex of the prisms (the second prism may be slightly off set vertically). Arrange the second set of prisms in the same manner (opposite vertical off set). This gives a "mirror" image of the rays in the first set and the recombination of the dispersed light.

30. In general if the sunlight is not blocked by cloud cover and the Sun is not over 42° altitude, conditions are such that a rainbow could be seen somewhere. Also, the Sun must be behind an observer to see a rainbow.

31. (a) Angular conditions for rainbow apply to a circular arc or cone. (b) Double internal reflection (not "total") and droplet population may be less at higher altitude.

32. The rainbow position or the "rainbow" seen depends on the relative position of the Sun and the observer, so different observers may see rainbows at different positions, e.g., observers at different elevations may see arcs of different elevations.

33. Because of the angles at which the rays are received, the red appears to be above the violet (see Fig. 24.17).

34. The cloud ice crystals act like hexagonal prisms that with refraction and reflection give a cone of transmitted light (halo) from appropriately oriented crystals. Principle is similar to rainbow formation.

35. A converging lens is thicker in the center of the lens and thin at the edges. Rays of light passing through the lens will converge at a point beyond the lens. Converging lenses produce both real and virtual images. A diverging lens is thicker at the edges and thin at the center of the lens. Rays of light passing through the lens diverge from a point in front of the lens. Only virtual images are formed by diverging lens.

36. The image of the Sun. The energy collected by the
     lens over its cross-sectional area is concentrated
     or focused into a small area giving a greater
     energy density (analogous to pressure changes with
     area).

37. The lens would form a parallel beam on the opposite
     side of the lens from the light source.

38. (a) Spherical aberration. (b) Possible chromatic
     aberration due to dispersion as light passes
     through glass of the mirror and distortion due to
     reflections from front and back surfaces.

SAMPLE TEST QUESTIONS

Multiple Choice

1. Ray optics is a convenient way to represent (a)
    interference, (b) diffraction, *(c) reflection, (d)
    wavelength.

2. In optics, the angle of incidence is (a) measured
    from the surface of a material, (b) equal to the
    angle of refraction, (c) is less than the angle of
    reflection for diffuse reflection, *(d) measured
    from a normal to the surface.

3. The "beam" of a flashlight is seen as a result of
    *(a) diffuse reflection, (b) refraction, (c)
    dispersion, (d) internal reflection.

4. A light beam obliquely incident on the side of a
    glass of water emerges from the other side. In
    passing through, the light is refracted how many
    times? (a) 2, (b) 3, *(c) 4, (d) 5, (e) 6.

5. Being able to "see hot air" rising from a hot
    surface is due to (a) diffuse reflection, (b)
    dispersion, *(c) refraction, (d) internal
    reflection.

6. The critical angle for a water-air interface is
    about 48°. Light will be transmitted from the
    water for an angle of incidence of

    (a) 60°, (b) 52°, (c) 48°, *(d) 44°

7. Fiber optics is based on (a) diffuse reflection,
    (b) dispersion, *(c) total internal reflection, (d)
    diverging mirrors.

8. Monochromatic light transmitted into a prism cannot be (a) refracted, (b) totally internally reflected, *(c) dispersed, (d) reflected by 180°.

9. Which of the following colors of light is refracted more in a typical transparent medium? *(a) blue, (b) yellow, (c) green, (d) red.

10. Dispersion is responsible for (a) diffuse reflection, (b) a diamond's brilliance, (c) spherical aberration, *(d) chromatic aberration.

11. Light rays (a) are always parallel, (b) can be circular, *(c) propagate perpendicular to wave fronts, (d) display the wave nature of light.

12. The law of reflection (a) applies only to plane mirrors, (b) is limited to regular reflection, (c) must be modified for diffuse reflection, *(d) applies to all reflecting surfaces.

13. An example of a diverging spherical mirror would be a (a) bathroom mirror, *(b) hubcap, (c) TV dish, (c) flashlight reflector.

14. An example of a converging spherical mirror would be a (a) spherical truck mirror, (b) spherical Christmas tree ornament, (c) bathroom mirror, *(d) reflector for a surgical or dentist's lamp.

15. Air is "seen" rising from a hot surface as a result of (a) reflection, *(b) refraction, (c) dispersion, (d) aberration.

16. When light enters a more optically dense medium at an angle, (a) $\theta_1 = \theta_2$, *(b) $\theta_1 > \theta_2$, (c) $\theta_1 < \theta_2$, (d) $\theta_1 > \theta_c$

17. When light is incident on the boundary of a less optically dense medium at an angle equal to the critical angle, *(a) $\theta_2 = 90^{\circ}$, (b) $\theta_1 = \theta_2$, (c) $\theta_1 > \theta_2$, (d) none of the preceding.

18. When light is refracted in a more optically dense medium, *(a) the frequency is unchanged, (b) the light speed increases, (c) $\theta_1 = \theta_2$, (d) the wavelength increases.

19. Stars twinkle as a result of atmospheric (a) dispersion, (b) reflection, *(c) refraction, (d) aberration.

20. Spherical aberration (a) is the same as chromatic aberration, (b) results from dispersion, (c) is a case of diffuse reflection, *(d) is not a problem for parabolic mirrors.

21. The property of light that exhibits it as a wave is (a) interference, (b) dispersion, (c) diffraction, *(d) all of these.

22. Adjacent portions of a wave that are in phase yield a form called (a) refraction, (b) dispersion, (c) a ray, *(d) a wave front.

23. A change in the direction of a wave due to a boundary is called (a) refraction, (b) interference, *(c) reflection, (d) polarization.

24. A change in the direction of a wave due to a change in its velocity is called *(a) refraction, (b) reflection, (c) interference, (d) dispersion.

25. Reflection from a plane mirrow is called _____ reflection. (a) diffuse, (b) regular, (c) specular, *(d) b and c.

26. A converging spherical mirror has a _____ surface that is reflecting.
    (a) convex, *(b) concave, (c) plane, (d) specular

27. The separation of sunlight into component colors by means of a prism is called (a) diffraction, *(b) dispersion, (c) refraction, (d) none of these.

28. Images of objects are formed by lenses due to (a) diffraction, (b) internal reflection, *(c) refraction, (d) all of these.

29. Spherical lenses may not form perfect images, owing to material defects and to various natural and inherent effect called (a) mass defects, *(b) aberrations, (c) dispersions, (d) all of these.

30. A lens thicker at the center than at the edge is called a _____ lens. (a) converging, (b) convex, (c) diverging, *(d) both a and b.

Completion

1. Reflection from a mirror surface is called <u>regular or specular</u> reflection.

2. Light passes from air into water. The speed of light <u>decreases</u> when it enters the water. $\sqrt{2}-3$

3. Light passes from air into water. The frequency of the light <u>remains the same</u>.

4. Light passes from air into water. The wavelength of the light <u>decreases</u> when it enters the water.

5. The effect due to the slightly different speeds of light of different colors of light in a medium is called <u>dispersion</u>.

6. Because of the converging property of a convex spherical lens, this type of lens is sometimes referred to as a <u>converging</u> lens.

7. If very little light is reflected from a surface, it appears to be <u>black</u>.

8. A diverging spherical mirror is <u>convex</u> shaped.

9. A light beam, e.g., from a flashlight, can be seen because of <u>irregular or diffuse</u> reflection.

10. The law of reflection states that the angle of incidence is <u>equal to</u> the angle of reflection.

11. In geometrical optics, light is represented as <u>rays</u>.

12. If light is incident on a mirror at an angle of 55° relative to the normal, the reflected ray will be

    at an angle of <u>55°</u>.

13. A mirror is any smooth surface that <u>regularly or specularly</u> reflects light.

14. A beam of light parallel to the axis of a converging spherical mirror converges at the <u>focal point</u> after reflection.

15. Mirrors used in stores for monitoring are <u>(diverging) convex</u>.

16. When light goes from air into water, the angle of incidence is <u>greater than</u> the angle of refraction.

17. The "wet spot" commonly seen on roads in the summer is a reflection of <u>sky light</u>.

18. Total internal reflection occurs when the angle of incidence exceeds the <u>critical angle</u> for the particular material.

19. When sunlight passes through a prism, a display of colors is seen as a result of <u>dispersion</u>.

20. A line drawn perpendicular to wave fronts that indicates the direction of wave propagation and represents the wave is called a/an <u>ray</u>.

21. The reflecting surface of a converging spherical mirror is <u>concave</u> in shape.

22. "Shafts" of sunlight through clouds or as seen in a forest are due to <u>diffuse</u> reflection.

23. Satellite TV "dishes" are <u>converging</u> mirrors for radio and TV waves.

24. When light enters a less optically dense medium, it is bent <u>away from</u> the normal.

25. Fiber optics depends on internal <u>reflection</u>.

26. Dispersion results from <u>refraction</u>.

27. A rainbow is seen when you look <u>away from</u> the Sun.

28. The properties of lenses depends on <u>refraction</u>.

51-4

29. If the angle of incidence is 30°, the angle between a reflected ray and the surface is <u> 60 </u> degrees.

30. When light passes from water into air, the angle of refraction is <u>greater</u> than the angle of incidence.

31. If the angle of incidence is greater than the <u>critical angle</u> of a medium, then the light is internally reflected.

32. An object is placed 50 cm in front of a plane mirror. The distance between the object and the image is then <u> 100 </u> cm.

33. The normal (line) is <u>perpendicular</u> to the surface at a particular point.

34. The law of reflection is valid for <u>any</u> reflecting surface.

35. Mirages are formed because of refraction or internal <u>reflection</u>.

36. The distortion of an image due to the failure of the light to converge at a point or in a plane is called <u>aberration</u>.

Chapter 21

Matching

(Choose the appropriate answer from the list on the right.)

_e_	1.	wave front	a. defect of focus
_g_	2.	ray	b. perpendicular to surface
_i_	3.	reflection	c. due to a change in velocity
_k_	4.	concave mirror	d. a lens that converges parallel light
_o_	5.	convex mirror	
_c_	6.	refraction	e. adjacent parts of wave that are in phase
_n_	7.	total internal reflection	f. where parallel rays converge after reflection from a concave mirror
_j_	8.	dispersion	g. perpendicular to wave front
_d_	9.	convex lens	h. optical illusion
_m_	10.	concave lens	i. specular or diffuse
_a_	11.	aberration	j. the separation of polychromatic light into its component wavelengths
_h_	12.	mirage	k. converges parallel light rays incident on its surface
_l_	13.	law of reflection	l. $\theta_1 = \theta_r$
_b_	14.	normal	m. diverges parallel light rays
_f_	15.	focal point	n. incident angle above critical angle
			o. diverges parallel light rays incident on its surface

202

Chapter 22 Vision and Optical Instruments

Answers to Questions

1. The surface of the cornea.

2. The camera shutter remains closed and opens only for film exposure or viewing, while the eyelid shutter of the eye is open for continuous viewing. The camera shutter remains open longer for time exposures.

3. Through another sense, touch.

4. The eye "camera" essentially has both black and white "film" (the rods) and color "film" (the cones).

5. This black-and-white vision when the light intensity is too low (e.g., at twilight) to activate the color sensitive cones.

6. Vision is blurred because the ciliary muscles cannot effect accommodation of the crystalline lens for focusing.

7. The answers should be "evident", particularly with measurements done on some of the figures.

8. The nearsighted correction would be done with a divergent lens, and the astigmatism correction depends on the nature of the shape of the crystalline lens, e.g., a greater corrective lens curvature (converging) would be used when the crystalline lens has deficient curvature.

9. As shown in Fig. 22.5, the refractive properties of the lens for the light rays given an enlarged image. (You may wish to explain the elements of ray diagrams to your students to illustrate image characteristics.)

10. Moonlight is generally not intense enough to activate the color sensitive cones.

11. In general we say that white (light) is the presence of all colors and black (no light) is the absence of all colors. However, we commonly say white and black are "colors." Keep in mind that color is highly subjective.

12. Lacking blue cones, a person would be able to see blue-violet colors by use of green cones, but could not distinguish among these because of lack of blue cones for contrast. Therefore, blue-green color blindness because of the problems of distinguishing colors of the shorter visible wavelengths.

13. (a) A white wall would appear to be magenta (purplish-red) in color. (b) A blue wall would appear blue (red absorbed, blue reflected). (c) A green wall would appear dark since it would absorb most of the red and blue light. Magenta is "minus green", so green is "minus magenta."

14. Wavelengths or color not at or near the filter color.

15. A cyan object appears dark since cyan pigment is "minus red." A yellow object appears slightly red (pink), since red is close enough to yellow to be reflected by the yellow "oscillators."

16. A red spot light on a dark-blue suit would make it appear dark. Spot lights of similar color(s) to the color of clothes would be reflected and the clothes would stand out with properly contrasted colored backgrounds.

17. (a) Green with yellow sunglasses since the yellow sunglasses transmit green light (close enough to yellow). (b) Green board would appear dark with rose-colored glasses. (c) The colors of the glasses.

18. (a) Red. (b) Blue. (c) Blue (blue oscillators close to green).

19. Black, since energy would be absorbed and not reflected.

20. To prevent scattering of stray light.

21. Because of greater blue component in light from fluorescent lamps.

22. Because of subtraction. For example, blue is the complement of yellow, but subtractively, yellow is "minus blue."

23. Because the atoms of a white object has a greater range of resonance frequencies and can respond to almost any color light.

24. Additive color method. Red and cyan are complementary colors.

25. In general not by the subtractive method with primary pigments, since components are absorbed. Some substances are white because they scatter all color components.

26. Additive. The color triads are stimulated to emit combinations of primary colors.

27. (a) Yellow + blue pigments produce green, i.e., green not absorbed. (b) Blue + red pigments produce yellow, i.e., yellow not absorbed.

28. In dense numbus clouds with larger water droplets, more light is absorbed than scattered, hence they appear dark.

29. Sun and skylight mask the light from the stars so they are not seen. On the moon, stars would be seen except in the vicinity of the bright Sun (or when blocked by Earth).

30. Due to Rayleigh scattering of increased shortened wavelength components in dense air near the earth at sunset. Longer wavelengths are scattered less and being transmitted to an observer give the characteristic color of beautiful sunsets. When low pressure air masses are to the west at sunset, the scattering is less and sunsets are not as colorful.

31. High pressure air masses are generally associated with fair weather, and when to the west give "red skys at night" due to Rayleigh scattering. "Red sky at morning" indicates a high pressure air mass to the east and a poor weather low pressure air mass follows from the west.

32. Yes, by using a second converging lens that would give a second inversion so the projected image would be upright.

33. By drawing slightly different pictures in time. The successive showing of these pictures would give the illusion of motion, similar to the frames of a "motion" picture. (You may recall old cartoon books that by flipping the pages gave animation.)

34. Up and down depend on the reference frame, e.g., think of "up and down" as the North pole and South pole of the earth. Also, astronomical studies are usually done with photographs that can be easily inverted.

35. (a) The light gathering ability of the lens or mirror limits the distance an object can be detected. (b) Reflecting telescopes have greater advantages over the refracting telescopes because of physical limitations on the refracting telescopes. Also, a reflecting telescope has only one surface to perfect, has less material defects, and aberration effects are easier to correct.

36. The image of the crime is inverted, and the field of view of the crime is small. As defense attorney, these two points should be emphasized.

37. (a) An Earth-orbiting telescope would not have to view stars and galaxies through the Earth's atmosphere, thus, a clearer view. (b) No.

38. (a) 200 inches equals 5.1 meters, (b) Area of mirror is .31416 in² or 218 ft² or 20.4 in².

39. Advantages: low operating cost; savings of natural resorces such as, coal, oil, and gas; reduces the pollution of the environment.
    Disadvantages: high cost of construction; variable weather prevents continuous use.

SAMPLE TEST QUESTIONS

Multiple Choice

1. The aperature through which light enters the eye is the (a) sclera, (b) cornea, *(c) pupil, (d) retina.

2. In the late evening, no color is seen because of lack of stimulation of (a) rods, *(b) cones, (c) cornea, (d) crystalline lens.

3. A nonspherical eyeball can give rise to (a) nearsightedness, *(b) astigmatism, (c) color blindness, (d) Rayleigh scattering.

4. Light with wavelengths greater than 600 nm appears to have the general color of (a) blue, (b) green, (c) yellow, *(d) red.

5. If light with wavelengths of 550 nm and 430 nm is seen by the eye, it would appear *(a) blue, (b) green, (c) orange, (d) red.

6. White is seen on a screen when green light is mixed with what color of light? (a) blue, (b) cyan, *(c) magenta, (d) ultraviolet. 53-3n 54-5

7. If turquoise and purplish-red paints (pigments) are mixed, the mixture appears (a) red, *(b) blue, (c) yellow, (d) orange.

8. The sky appears blue as a result of (a) selective absorption, (b) selective reflection, (c) selective transmission, *(d) preferential scattering.

9. A reflecting telescope uses (a) converging lens, (b) diverging mirror, *(c) converging mirror, (d) diverging lens.

10. A refracting telescope uses *(a) converging lenses, (b) diverging mirrors, (c) converging mirrors, (d) diverging lenses.

11. The amount of light entering the open eye is regulated by the (a) crystalline lens, (b) eyelid, *(c) pupil, (d) retina. 54-6

12. The film of a camera corresponds to what part of the eye? (a) aperture, *(b) retina, (c) cornea, (d) iris. 54-6

13. Persons that can see distant objects clearly but not close objects are *(a) farsighted, (b) nearsighted, (c) astimatic, (d) color blind.

14. Color vision (a) is affected by a receding near point, (b) is absent in color blindness, (c) may be corrected with reading glasses, *(d) none of the preceding.

15. Which of the following is not a primary color of light? (a) red, (b) green, *(c) yellow, (d) blue 53-3 54-5

16. The overlapping of beams of red and green light on a screen gives rise to the color of *(a) yellow, (b) cyan, (c) blue, (d) none of the preceding.

17. If white light is transmitted through a red filter and this light through a blue filter, the emerging light would appear, (a) red, (b) blue, (c) cyan, (d) black.

18. Rayleigh scattering explains (a) color vision, *(b) the blueness of the sky, (c) additive color production, (d) subtractive color production.

19. A refracting telescope (a) is always an astronomical telescope, (b) uses a converging mirror as an objective, (c) is the principle of the radio telescope, *(d) none of the preceding.

20. Various types of telescope, use (a) visible light, (b) radio waves, (c) infrared radiation, *(d) all of the preceding.

21. The changing of the curvature of the crystalline lens of the eye so that the images of objects at various distances are formed on the retina is called (a) astigmatism, (b) focal adjustment, *(c) accommodation, (d) radius of curvature.

22. Photosensitive cells of the retina that are responsible for twilight (black-and-white) vision are called *(a) rods, (b) cones, (c) meniscus, (d) sclera.

23. Photosensitive cells of the retina that are responsible for color vision are called (a) rods, *(b) cones, (c) stems, (d) pupils.

24. A vision defect in which a person can see near objects clearly, but not distant objects is known as (a) farsightedness, *(b) nearsightedness, (c) astigmatism, (d) hyperopia.

25. A vision defect in which a person can see far objects clearly, but not near or close objects is known as *(a) farsightedness, (b) nearsightedness, (c) myopia, (d) astigmatism.

26. A vision defect caused by the cornea and/or the crystalline lens not being spherical, so the image on the retina is out of focus in particular planes or directions is known as (a) hyperopia, (b) myopia, *(c) astigmatism, (d) amblyopia.

27. The additive primary colors are *(a) red, green, and blue, (b) red and blue, (c) red, yellow, and blue, (d) white and black.

28. The subtractive primary colors are (a) yellow, green, and blue, *(b) yellow, cyan, and magenta, (c) cyan and magenta

29. A cyan filter absorbs its complement _____ light.
*(a) red, (b) blue, (c) green, (d) yellow. ~~Same~~

30. A magenta filter absorbs its complement _____ light.
(a) red, (b) blue, *(c) green, (d) yellow.

Completion

1. Light enters into the interior of the eye through a hole called the <u>pupil</u>.

2. When the image is not formed on the surface of the retina of the eye, a person sees a <u>blurred or out-of focused image.</u>

3. The photosensitive cells in the retina of the eye are rods, which are responsible for <u>black and white or twilight</u> vision; and cones, which are responsible for <u>color</u> vision.

4. The sky appears blue because of <u>(Rayleigh) scattering</u>.

5. Two colors mixed together that produce white are called <u>complementary</u> colors.

6. The two basic lenses of a compound microscope are the <u>eyepiece or ocular</u> and the <u>objective</u> lenses.

7. The range of wavelength for visible light is about from <u>4000</u> to <u>7000</u> angstroms.

8. Color blindness occurs when <u>one type of primary cone</u> is missing.

9. The changing of the curvature of the crystalline lens of the eye so that images are formed on the retina is called <u>accommodation</u>.

10. The mixing of paints is example of the <u>subtractive</u> method of color mixing.

11. The cells of the retina responsible for black-and-white vision are called <u>rods</u>.

12. The loss of accommodation with age gives rise to a condition of <u>farsightedness</u> in older people.

13. A person that is red-green color blind is missing <u>red</u> or <u>green</u> cones.

14. Combined green and blue light appears cyan in color. Cyan and red light combined appears <u>white</u>.

15. The subtractive primaries are cyan, yellow, and magenta.

16. The gases of the atmosphere preferentially scatter light in the blue end of the visible spectrum.

17. Red sunsets require the air to contain particles.

18. The sound track of a motion picture movie is on the film in variations of optical density.

19. The largest astonomical telescopes are reflecting telescopes.

20. A Galilean telescope has a diverging lens for an eyepiece.

21. The "film" on the back surface of the eyeball on which images are formed and transmitted to the brain is called the retina.

22. The crystalline lens of the eye is a converging lens which focuses incoming light on the retina.

23. A refracting telescope has a converging lens as a collector.

24. To correct farsightedness where the image is formed behind the retina, glasses with converging lenses is used to bring the image to focus on the retina.

25. Parabolic mirrors are used in high quality reflecting telescopes so as to avoid spherical aberration.

26. A nearsighted person cannot see far objects clearly.

27. The eye is most sensitive to yellow-green light.

28. Red, blue, and green are the primary colors.

29. The combined lights of complementary colors appear white.

30. Paint mixing to produce desired colors is an example of the subtractive method of color production.

31. Rayleigh scattering causes the sky to appear blue.

32. A simple microscope consists of a single converging lens.

33. A terrestrial telescope has an <u>upright</u> image.

34. Parabolic mirrors are used in high quality reflecting telescopes so as to avoid spherical <u>aberration</u>.

35. A yellow filter absorbs its complement <u>blue</u> light.

36. The additive mixture of red, green, and blue light produces <u>white</u> light.

37. The subtractive combination of cyan, magenta, and yellow filters transmits <u>no</u> light.

38. The color perceived for monochromatic light depends on the <u>frequency</u> of the light.

39. Any two colors which combine to form white light are said to be <u>complementary</u>.

40. A piece of red cloth would appear <u>black</u> when illuminated by green light in a closed room.

Chapter 22

Matching

(Choose the appropriate answer from the list on the right.)

<u>d</u> 1. retina

<u>m</u> 2. accommodation

<u>o</u> 3. rods

<u>f</u> 4. cones

<u>l</u> 5. nearsightedness

<u>b</u> 6. farsightedness

<u>j</u> 7. astigmatism

<u>c</u> 8. primary colors

<u>h</u> 9. complementary colors

<u>a</u> 10. subractive primaries

<u>p</u> 11. Rayleigh scattering

<u>n</u> 12. refracting telescope

<u>e</u> 13. reflecting telescope

<u>i</u> 14. the combination of blue and green light

<u>k</u> 15. the color scattered least in the atmosphere

<u>g</u> 16. the complement of blue light

a. cyan, magenta, and yellow

b. receding near-point

c. red, blue, and green

d. the photosensitive part of the eye

e. mirror

f. responsible for color vision

g. yellow

h. combinations of colors that appear white to the eye

i. cyan

j. a vision defect when images on retina is out of focus in particular planes

k. red

l. capable of seeing near objects

m. changing of the curvature of the crystalline lens

n. lens

o. responsible for twilight vision

p. blueness of sky

Chapter 23   Relativity

Answers to Questions

1. (a) about 8 minutes (from the Sun), (b) 4.2 light years (from Alpha Centauri).

2. $d = ct = (3 \times 10^8$ m/s$)(0.2$ s$) = 0.6 \times 10^8$ m $=$

    6×10^4 km.   (60,000 km or 37,000 mi)

3. (a) A translucent object would contain only a small amount of ether (little light transmission), and (b) an opaque object would contain little or no ether (no light transmission).

4. The light from stars began its journey in the past (often many years ago), so the light received on Earth is old or a view into the past.

5. The contraction would be too small to measure by ordinary methods, and optically by inference methods the contraction is assumed to occur to explain the results.

6. No, taking into account the motion of the source, the stationary observer would say the frequency of the source was 1000 Hz (Doppler effect). Also, if the stationary observer sounded a 1000-Hz horn, the moving observer would hear a 1010 Hz sound.

7. Yes, but ordinary speeds are much less than c, so there is no observed relativistic effects. (0).

8. The observer in car B sees car A not moving, and sees car C moving away with a speed of 10 km/h. The observer in car C sees car A and car B slowing down with a speed of 10 km/h.

9. Because it's unitless, i.e., $(v/c)^2$ is a ratio in which the units cancel.

10. Each reference frame has its own proper time. Observing the program on a TV set in the moving frame, an observer would see the program running behind (time dilation) the program on his TV set.

11. For $L = L_o/2 = L_o/\gamma$, then $\gamma = 2$. The mass would be twice as great since $m = \gamma m_o = 2m_o$.

12. The length of the meterstick would remain the same, i.e., 1 m, (but a contraction in its width in the same direction of motion) and the mass "dilation" would still be by a factor of 2 ($m = 2m_o$).

13. Same height (1.8 m), but thinner.

14. Twin paradox. (a) An Earth observer would think so with the proper time dilation (very fast spaceship) since the observed time on the spaceship would be slower. (b) The astronaut might have some doubts referencing time to his clock.

15. You would observe your (a) pulse rate, (b) mass, and (c) volume to be normal ("proper" frame quantities).

16. (a) Since a child is always younger than its parents, a child astronaut would return with an even greater age difference with the parents. (b) In this case, it might be possible for the returning parents to be younger than their child.

17. Appreciable mass is converted into energy only in certain processes, e.g., nuclear fission.

18. An outside observer would see the ball traveling in a straight line according to Newton's first law with the spaceship accelerating upward so the floor hits the ball.

19. In one second light travels 3×10^8 m/s or about 186,000 mi, so it can't be observed for one second in a uniform gravitational field as for the ball. (Even for a fraction of a second the effect would be unnoticed.)

20. Any deflection of the laser beam due to the Earth's gravitational field is negligible over such short distances.

21. In the rotating system clock C is accelerating toward clock A (centripetal acceleration) and hence runs slower.

SAMPLE TEST QUESTIONS

<u>Multiple Choice</u>

1. An unsuccessful attempt to measure the speed of light by *(a) Galielo, (b) Fizeau, (c) Einstein, (d) Fitzgerald.

2. The existence of ether (a) is necessary for light propagation, (b) was determined by the Michelson-Morely experiment, *(c) would provide an absolute reference, (d) is postulated in the principle of relativity.

3. A postulate of the special theory of relativity is (a) the existence of ether, *(b) the constancy of the speed of light, (c) the Fitzgerald contraction, (d) the principle of equivalence.

4. A beam of light emitted in a system moving at a speed of c/2 relative to an observer would be measured by the observer to have a speed of (a) -c/2, *(b) c, (c) c+, (d) c - (c/2).

5. The principle of relativity deals with (a) ether, (b) speed of light, (c) mass and energy, *(d) laws of physics.

6. A relativistic time dilation gives rise to (a) ether effects, (b) mass-energy, *(c) the twin paradox, (d) the principle of equivalence.

7. A measured time dilation and length contraction would affect the computed speed by a factor of

(a) γ, *(b) $1/\gamma^2$ (c) γ^{-1}, (d) 1 (no effect).

8. A postulate of the general theory of relativity is the *(a) principle of equivalence, (b) principle of relativity, (c) constancy of the speed of light, (d) none of the preceding.

9. A result of the general theory is that gravity causes (a) a length contraction, *(b) time to slow down, (c) the ether drag, (d) a Fitzgerald contraction.

10. An astronaut inside a closed spaceship believes that he is in free space as a result of experiments. However, the experimental results could also be due to (a) a length contraction, (b) a time dilation, (c) the constancy of the speed of light, *(d) an acceleration.

11. All laws of physics are the same for all observers moving at a constant velocity with respect to one another. This is known as the (a) special theory of relativity, *(b) principle of relativity, (c) general theory of relativity, (d) principle of equivalence.

12. Einstein's theory that deals with non-accelerating systems is called the _____ theory of relativity. *(a) special, (b) general, (c) equivalence, (d) principle.

13. The Michelson-Morely experiment was designed to (a) determine the velocity of light *(b) detect the ether, (c) prove Einstein's theory of relativity, (d) all of the above.

14. An inertial system refers to (a) a nonaccelerating system, (b) a constant velocity system, (c) a system at rest, *(d) all of these.

15. Einstein's theory of relativity that deals with accelerated systems is called the (a) special theory, (b) principle of equivalence, *(c) general theory, (d) principle of relativity.

16. In respect to the development of the theory of relativity, the term ether refers to (a) the solar wind, *(b) a hypothetical medium for the propagation of light waves, (c) an anesthetic, (d) empty space.

17. The special theory of relativity is based on (a) the principle of relativity, (b) the constant speed of light, *(c) both a and b.

18. The principle of equivalence refers to the observed effects resulting from either (a) acceleration or inertial mass, (b) acceleration or velocity, *(c) acceleration or gravity, (d) mass or energy.

19. The lifetime of a muon, that is in motion relative to an observer, is (a) the same as its lifetime at rest, *(b) greater than its lifetime at rest, (c) less than its lifetime at rest, (d) none of the above.

20. A contraction or shortening of the length of the arm of the Michelson-Morley interferometer in the direction of the ether wind was proposed by (a) Michelson, (b) Morley, (c) Einstein, *(d) Fitzgerald.

21. An observer looking at a meterstick in another
    moving inertial system observes it (a) to be less
    than one meter when the stick is in a direction
    perpendicular to the direction of motion, (b) to be
    its proper length, *(c) to have smaller centimeter
    intervals than on his meterstick when the stick in
    in a direction parallel to the direction of motion,
    (d) to have a dilation.

22. When an object travels at a speed that is an
    appreciable fraction of the speed of light, it is
    observed to have (a) a size increase, *(b) a mass
    increase, (c) a length increase, (d) no changes.

23. According to the general theory of relativity, (a)
    light is unaffected by a gravitational field, (b) a
    twin would return from a space journey older than
    his twin on Earth, *(c) you wouldn't be able to
    tell if you were in an accelerated system or a
    gravitational field, (d) inertial systems are
    different.

Completion

1. A spaceship travels with a speed of c/2. A light
    shines in the direction of motion of the ship.
    From another reference frame the speed of light
    would be measured to be __c__.

2. As the speed of an object increases, its
    relativistic mass __increases__.

3. Einstein's second postulate of the special theory
    of relativity explained the results of the
    __Michelson-Morely__ experiment.

4. A device used to measure small differences in wave
    interference is called __an interferometer__.

5. A particle resulting from cosmic ray collisions
    with gas molecules of the air and gives
    experimental evidence of relativistic time dilation
    is the __muon__.

6. A system at rest or in uniform motion is called __an
    inertial__ system.

7. A light year is the distances __light travels in one
    year__.

8. In the equation $E = mc^2$, the E stands for __energy__,
    the m for __mass__, and c is the __speed of light__.

Chapter 23

9. In the length contraction, a length is observed to contract in the <u>same</u> direction as the motion.

10. If a body could travel at the speed of light, according to the theory of relativity its mass would be <u>infinite</u>.

11. The light year is a unit of <u>distance</u>.

12. The Michelson-Morely experiment was based on detecting the velocity addition properties of light and the <u>ether wind</u>.

13. The instrument developed by Michelson to detect the ether was called <u>an interferometer</u>.

14. According to the <u>principle of relativity</u> all the laws of physics are the same for all inertial observers.

15. Relative motion gives rise to a length <u>contraction</u> and a time <u>dilation</u>.

16. Experimental evidence for time dilation is provided by <u>muon</u> decay.

17. Relativistic mass increases are observed in particle <u>accelerators</u>.

18. At normal speeds, the relativistic equations reduce to <u>classical</u> measurements.

19. The principle of equivalence states that one cannot distinguish between a <u>gravitational field</u> and the effects of <u>an acceleration</u>.

20. Evidence for the gravitational bending light is obtained at the time of <u>a solar eclipse</u>.

21. According to the twin paradox, the twin in the spaceship will return <u>younger</u> than the twin who remains on Earth.

22. In a spaceship accelerating at a rate of 9.8 m/s² in space, the effects would be <u>the same</u> as the Earth's surface.

23. The gravitational force of a black hold would cause light passing nearby to be bent <u>toward</u> the black hole.

24. The gravitational red-shift causes light to have <u>smaller</u> frequencies.

25. According to Einstein's theories, the greatest speed is that of the speed of light in <u>vacuum</u>.

26. When mass is converted to energy, it is generally in the form of <u>heat (kinetic energy)</u> and light.

27. In the Sun, mass is converted to <u>energy</u>.

28. The <u>special</u> theory of relativity is limited to <u>inertial</u> systems.

29. Prior to Einstein, ether was believed to be the material for <u>light</u> propagation through space.

30. The speed of light was first successfully measured on Earth by <u>Fizeau</u>.

31. The speed of light in free space is about <u>300,000</u> km/s.

32. The Michelson-Morely experiment did not establish the existence of the <u>ether</u>.

33. Einstein received a Nobel Prize for his theory of <u>the photoelectric effect</u>.

34. According to the principle of <u>relativity</u>, you can not tell by experiment whether you are moving uniformly or are at rest.

35. A time dilation means that a moving clock appears to run <u>slower</u> to a "stationary" observer.

36. The principle of equivalence is a postulate of the <u>general</u> theory of relativity.

37. Since a light beam is bent in an accelerated system, the general theory of relativity predicts that the same would occur in a <u>gravitational field</u>.

38. According the the special theory of relativity the speed of light in free space is <u>constant</u> for all observers regardless of the motion of the source or the motion of the observer.

Chapter 23

Matching

(Choose the appropriate answer from the list on the right.)

___d___ 1. ether

___g___ 2. Michelson-Morley

___i___ 3. Fitzgerald

___n___ 4. special theory of relativity

___f___ 5. principle of relativity

___l___ 6. velocity of light

___b___ 7. time dilation

___o___ 8. length contraction

___j___ 9. proper time and length

___c___ 10. twin paradox

___m___ 11. relativistic mass

___e___ 12. mass-energy conversion

___a___ 13. general theory of relativity

___k___ 14. principle of equivalence

___h___ 15. Fizeau

a. deals with accelerated systems

b. proper value times

c. predicted by the general theory of relativity

d. hypothetical medium

e. $E = mc^2$

f. laws of physics are same for observer moving at constant velocity

g. designed an experiment to detect motion of Earth through the ether

h. measured speed of light using cogwheel

i. proposed contraction of the interferometer arm

j. measurement made by an observer moving with the clock and object

k. accelerated system versus gravitational system

l. constant in free space

m. $m_o \gamma$

n. deals with non-accelerating systems

o. proper value divided by γ

Chapter 24   Quantum Physics

Answers to Questions

1. The color is indicative of the emitted spectral line of maximum intensity. This is in the red region at moderately hot temperatures.

2. Planck's quantum theory gave an explanation for the frequency distribution of thermal radiation.

3. (a) Blue light since it has a greater frequency. (b) Visible light quanta since they have greater frequencies. (Red end of spectrum has longer wavelengths and lower frequencies, $f = c/\lambda$.)

4. Because there are so many quanta that their discrete nature cannot be detected by a radio receiver.

5. The cutoff frequency is below that of visible light or at least most of the frequencies of visible light.

6. The meter could be broken, but this is easily checked. More likely, the cutoff frequency of the meter photocell is above that of red light.

7. The amount of water in the incident bucket (photon with energy hf) is less than that required to rill the other bucket (work function of Photoelectron, hf_o).

8. Not really. White light is made up of frequency (color) components and hence quanta of different energies.

9. Ultraviolet light has a higher frequency than the components of visible light and hence is more energetic ($E = hf$).

10. The light beam (which may not be in the visible region for a burglar alarm) and the photocell form part of a circuit. When the beam is interrupted, the current in the circuit is interrupted, which with the proper circuitry activites a counter or causes an alarm to sound.

11. By spectral analysis of sunlight, i.e., comparing the solar spectrum with the spectra of known elements.

12. In solids, the atoms are closely packed and interactions influence the electron states to the point that an electron from one atom can jump to the level of another atom. Excitation occurs randomly and the atoms emit randomly at different frequencies, giving a continuous spectrum.

13. No, a discrete spectrum since the electrons in the atoms of an excited gas undergo certain transitions.

14. The absorbed light is scattered or re-emitted in all directions by the atoms, so light is scattered from the transmission direction giving dark lines.

15. Centripetal acceleration due to a change in velocity (change in direction).

16. Because it is (centripetally) accelerated and classically an accelerated charged particle emits energy (radiation), which would cause the electron to spiral into the nucleus.

17. $n = 3$ to $n = 1$, since there is a greater difference in the energy states.

18. Six (if all transitions are allowed). For $(n_i - n_f)$, there are the transitions $(4 - 3)$,

    $(4 - 2)$, $(4 - 1)$, $(3 - 2)$, $(3 - 1)$, and $(2 - 1)$.

19. Centripetal forces in both cases, but gravitational versus electrical. A satellite could be put into an orbit of any radius, but a hydrogen electron must be in orbits with discrete or particular radii. Also, if a satellite loses energy (e.g., due to frictional losses), it would spiral into the Earth (if not burned up in the atmosphere).

20. Randomly or chaotically for the incoherent case. If the people marched in-step with the same cadence in lines (so as not to interfere or in phase), such as a marching band, this is analogous to being coherent.

21. So there can be a net emission of photons. Without an inverted population, and more atoms in the ground state than in excited states, there would be a net absorption of energy and the laser would have no output.

22. No, in that there would be no amplification if the light escaped through the glass end of the tube. There must be appreciable reflection to give amplification by stimulated emission and to set up the beam along the tube axis.

23. No, light from an incandescent source is incoherent.

24. Exactness versus probability.

25. Yes. A beam of electrons produces a diffraction pattern.

26. Assuming a 140 lb person, the mass is approximately 140 lb (0.5 kg/lb) = 70 kg; and a running speed of about 22 mi/h or 36 km/h $(10^3$ m/km)(1 h/3.6 x 10^3 s) = 10 m/s. Then, λ = h/mv = (10^{-34})/(70)(10) = 1.4 x 10^{-37} m.

27. Yes. The measurement process effects the quantity being measured. It is physically impossible to simultaneously know a particle's exact position and velocity.

28. Δ x Δ p $\approx$ h, therefore the velocity measurement would be reduced by 50 percent.

SAMPLE TEST QUESTIONS

Multiple Choice

1. When the temperature of an incandescent solid is increased, (a) the emitted light intensity is less, (b) there is an ultraviolet catastrophe, *(c) the most intense spectral component is shifted to a higher frequency, (d) nothing changes.

2. The quantum hypothesis was introduced by (a) Einstein, *(b) Planck, (c) Bohr, (d) deBroglie.

3. In the photoelectric effect, *(a) the photocurrent is proportional to the light intensity, (b) the electron kinetic energy is independent of the light frequency, (c) the electron kinetic energy is dependent on the light intensity, (d) photoemission occurs below the cut-off frequency for high light intensities.

4. The greater the frequency of the light in the photoelectric effect, (a) the less the time delay for photoemission, *(b) the greater the kinetic energy of the photoelectrons, (c) the greater the light intensity, (d) the lower the threshold frequency.

5. The photoelectric effect was explained by (a) Planck, *(b) Einstein, (c) Bohr, (d) deBroglie.

6. The work function of a photoelectric material, (a) is always equal to the kinetic energy of the photoelectron, (b) is given by a principal quantum number, (c) depends on the light intensity, *(d) is the energy needed to free an electron.

7. When the principal quantum number is one, the hydrogen electron is (a) in an excited state, (b) ionized from the atom, *(c) in the ground state, (d) going to decay to a lower energy level in a short time.

8. When a hydrogen electron is excited to a higher energy level (a) it must be in the ground state, (b) a photon of any frequency may be absorbed, (c) a photon must be emitted, *(d) energy is absorbed.

9. Light amplification in a laser is due to (a) spontaneous emission, (b) the photoelectric effect, *(c) stimulated emission, (d) the ultraviolet catastrophe.

10. Experimental evidence of deBroglie's hypothesis was demonstrated by (a) Einstein, *(b) Davisson and Germer, (c) Bohr, (d) Planck.

11. Planck's hypothesis applied to (a) lasers, (b) line spectra, (c) photons, *(d) thermal oscillators.

12. In the photoemission of electrons, the kinetic energy of the electrons depends on the light *(a) frequency, (b) intensity, (c) phase, (d) coherence.

13. The cut-off frequency of the photoelectric effect depends on (a) light intensity, (b) light frequency, *(c) material work function, (d) the dual nature of light.

14. The theory of the hydrogen atom was developed by (a) Planck, (b) Einstein, (c) deBroglie, *(d) *(d) Bohr.

15. The photon with the greatest frequency would be emitted for which of the following hydrogen atom transitions? (a) n = 1 to n = 2, *(b) n = 2 to n = 1, (c) n = 3 to n = 2, (d) n = 4 to n = 3.

16. The laser (a) produces incoherent light, (b) depends on spontaneous emission, *(c) is an optical laser, (d) was invented by Einstein.

17. Holograms depend on what property of laser light? *(a) coherence, (b) intensity, (c) frequency, (d) color.

18. The idea that particles are piloted or guided by waves was developed by (a) Planck, *(b) de Broglie, (c) Einstein, (d) Bohr.

19. Which of the following was not involved in producing experimental results that supported deBroglie's hypothesis? (a) Davisson, (b) Thomson, *(c) Townes, (d) Germer

20.. Matter waves are employed in which of the following applications? *(a) electron microscope, (b) laser, (c) code-bar scanning system, (d) smoke alarm

21. For an incandescent solid, the relative brightness or intensity of the different frequencies, which we see as colors, depends on the (a) wavelength, *(b) temperature, (c) pressure, (d) electron emission

22. Classical wave theory predicts that the intensity of the radiation from an incandescent solid should be proportional to the (a) frequency, (b) wavelength, *(c) frequency squared, (d) none of these.

23. Planck's hypothesis states that the radiated energy from a thermal radiator is (a) continuous, *(b) quantized, (c) inversely proportional to the frequency, (d) directly proportional to the frequency squared.

24. The emission of electrons from a material that is exposed to light is known as the _____ effect. (a) particle, (b) electromagnetic, *(c) photoelectric, (d) light

25. The dual nature of light refers to the (a) quantum properties of light, *(b) wave and particle characteristic properties of light, (c) frequency and wavelength properties of light, (d) none of these.

26. The most important feature that Bohr assumed in his theory of the hydrogen atom was the quantization of the _____ of the electron. (a) charge, (b) momentum, *(c) angular momentum, (d) electrical potential energy.

27. All particles in motion have wave properties. The waves that exhibit such properties are called _____ waves. (a) matter, (b) de Broglie, *(c) both a and b, (d) none of these.

28. When a transition of an electron takes place in an atom from an energy level of n = 3 to n = 2, (a) a discrete amount of energy will be absorbed, *(b) a discrete amount of energy will be emitted, (c) the atom becomes less stable, (d) none of the above.

29. In a stimulated emission process, an excited atom is struck by a photon of the same energy of the allowed transition, and the atom emits an identical photon. After emission, the two photons (a) have the same frequency, (b) are emitted in the same direction, (c) are in phase, *(d) all of these.

30. Laser light is (a) chaotic, (b) incoherent, *(c) coherent, (d) polychromatic.

Completion

1. Classically, it is predicted that the intensity of emitted thermal radiation should be proportional to the (square of the) frequency.

2. In the photoelectric effect, the photocurrent is proportional to the intensity of the light.

3. The color of a radiating body depends on its temperature.

4. The total number of bright lines in a spectrum which can be produced with three excited states above the ground state is 6 .

5. Planck's hypothesis states that the energy of an oscillator is proportional to its frequency.

6. The dual nature of light means that light sometimes behaves as a wave and sometimes as a particle.

7. In the Bohr theory, light is emitted as photons when an atomic electron goes from a higher energy state to a lower energy state.

8. The term laser is an acronym for <u>light amplification by stimulated emission of radiation</u>.

9. Two types of photo emission are <u>spontaneous</u> emission and <u>stimulated</u> emission.

10. In quantum mechanics, matter waves give an indication of the <u>probability</u> of the location of a particle.

11. The energy of a photon is equal to <u>hf</u>.

12. The greater the <u>intensity</u> of light, the greater the photocurrent in the photoelectric effect.

13. The kinetic energy of photoelectrons depends on the light <u>frequency</u>.

14. The light from a gas discharge tube has a <u>line</u> spectrum.

15. An excited state has a quantum number of <u>two</u> or greater.

16. The visible line spectrum of hydrogen is known as the <u>Balmer</u> series.

17. Stimulated emission from an excited atom is induced by <u>a photon</u>.

18. Light with waves of different frequencies and phases is said to be <u>incoherent</u>.

19. The wavelengths of de Broglie waves for particles are <u>shorter</u> the greater the mass of the particle.

20. Quantum mechanics deals with <u>probability</u> rather than exactness.

21. In an electron microscope a <u>magnetic</u> lens may be used to focus the beam of electrons.

22. It is physically impossible to simultaneously know (measure) a particle's exact position and velocity. This is known as <u>Heisenberg's uncertainty principle</u>.

23. Wave theory predictions for the thermal emission of light give rise to what is called the <u>ultraviolet</u> catastrophe.

24. If the energy of the light photons is <u>greater</u> than the work function of a photoelectric material, then there is photomission.

25. An electron in an excited state is <u>farther from</u> the nucleus than when in the ground state.

26. The frequencies of a bright-line spectrum for an element correspond with the same <u>frequencies</u> of the dark lines of its absorption spectrum.

27. <u>Stimulated</u> emission occurs when an excited atom is struck by a photon of the same energy of an allowed transition and the atom emits an identical photon.

28. Light can be amplified by using energy stored in <u>atoms</u>.

29. The ultraviolet catastrophe was a "catastrophe" because of <u>energy</u> considerations.

30. According to Planck's hypothesis, the energy of an oscillator is <u>quantized</u>.

31. The kinetic energy of a photoelectron is dependent on the <u>frequency</u> of the light.

32. The threshold frequency for photoemission is the same as the <u>cut-off</u> frequency.

33. The electron orbits in the Bohr theory are designated by <u>principal</u> quantum numbers.

34. Light is emitted from an atom when an electron makes a transition from a <u>lower</u> to a <u>higher</u> energy level.

35. The operation of a laser depends on <u>stimulated</u> emission.

36. The output of a laser is <u>coherent</u> light.

37. The de Broglie wave for an electron is a <u>photon</u>.

38. A beam of electrons can be defracted like <u>waves</u>.

Matching

(Choose the appropriate answer from the list on the right.)

__d__ 1. ultraviolet catastrophe

__g__ 2. quantum hypothesis

__o__ 3. .photoelectric effect

__n__ 4. photon

__i__ 5. dual nature of light

__k__ 6. hydrogen atom theory

__c__ 7. laser

__f__ 8. matter waves

__a__ 9. quantum mechanics

__l__ 10. ground state

__h__ 11. coherent

__e__ 12. incoherent

__p__ 13. work function

__b__ 14. uncertainty principle

__m__ 15. cut-off frequency

__j__ 16. holography

a. probability

b. Heisenberg

c. stimulated emission

d. intensity is proportional to the frequency squared

e. chaotic

f. de Broglie

g. Planck

h. light waves with the same frequency, phase, and direction

i. particle and wave

j. three dimensional images

k. Bohr

l. n = 1

m. lowest frequency for photoemission

n. quanta of energy

o. Einstein

p. amount of energy needed to free an electron from a photomaterial

Answers to Questions

1.   No, not with only one number. Different isotopes
     (nuclides) of the same element have the same proton
     number, and different elements may have the same
     mass numbers or the same neutron numbers.

2.   (a) $_{1}^{1}H_{0}$, $_{1}^{2}D_{1}$, and $_{1}^{3}T_{2}$. (b) HOH (or commonly H_2O),

     HOD, HOT, DOD, DOT, and TOT, where the water with
     tritium would be radioactive.

3. $_{50}^{114}Sn_{64}$, $_{50}^{115}Sn_{65}$, $_{50}^{116}Sn_{66}$, $_{50}^{117}Sn_{67}$, $_{50}^{118}Sn_{68}$,

     $_{50}^{119}Sn_{69}$, and $_{50}^{120}Sn_{70}$.

4.   The gravitational force is always attractive and
     would contribute (negligibly) to stability between
     nucleons but not the attractive force of the
     electrons on the nucleons. The repulsive
     electrical forces among nuclear protons would
     contribute to instability as would the attractive
     electrical forces among nuclear protons and orbital
     electrons. Also, there are repulsive electrical
     forces among the orbital electrons (with the
     exception of hydrogen).

5.   Since the hydrogen atom consists of only one proton
     and one electron, the uranium atom has 92 protons,
     146 neutrons, and 92 electrons, the uranium atom is
     slightly more than 238 times as massive as the
     hydrogen atom. The volume of each is the same,
     since their diameters are equal. Thus, the uranium
     density is approximately 238 times as great.

6. $_{3}^{6}Li_{3}$, $_{9}^{19}F_{10}$, $_{14}^{28}Si_{14}$, $_{22}^{48}Ti_{26}$, $_{35}^{80}Br_{55}$, $_{78}^{179}Pt_{101}$.

7. $_{83}^{209}Bi_{126}$. The attractive nuclear forces are

     short-range and the repulsive electrical forces are
     long-range.

8.   Perhaps because it emitted alpha "rays."

9.   An electric field could be used. The positive
     alpha particles would be deflected in the direction
     of the field and the negative beta particles in the
     opposite direction. Gamma rays are not deflected

10. (a) $^{226}_{88}Ra \longrightarrow {}^{222}_{86}Rn + {}^{4}_{2}He$, (b) $^{60}_{27}Co \longrightarrow {}^{60}_{28}Ni + {}^{0}_{-1}e$,

(c) $^{210}_{84}Po^* \longrightarrow {}^{210}_{84}Po +$.

11. $^{131}_{53}I \longrightarrow {}^{131}_{54}Xe^* + {}^{0}_{-1}e$ and $^{131}_{54}Xe^* \longrightarrow {}^{131}_{54}Xe +$.

12. $^{237}_{93}Np \longrightarrow {}^{233}_{91}Pa \longrightarrow {}^{233}_{92}U \longrightarrow {}^{229}_{90}Th \longrightarrow {}^{225}_{88}Ra \longrightarrow$

$^{225}_{89}Ac + {}^{221}_{87}Fr \longrightarrow {}^{217}_{85}At \longrightarrow {}^{213}_{85}Bi (\longrightarrow {}^{209}_{81}Tl)$ or

$(\longrightarrow {}^{213}_{84}Po) \longrightarrow {}^{209}_{82}Pb \longrightarrow {}^{209}_{83}Bi$

13. $^{233}_{91}Pa \longrightarrow {}^{233}_{92}U + {}^{0}_{-1}e$, (b) $^{233}_{92}U \longrightarrow {}^{229}_{90}Th + {}^{4}_{2}He$

14. (a) $^{222}_{86}Rn \longrightarrow {}^{218}_{84}Po + {}^{4}_{2}He$,

(b) $^{218}_{84}Po \longrightarrow {}^{214}_{82}Pb + {}^{4}_{2}He$ and

$^{218}_{84}Po \longrightarrow {}^{218}_{85}At + {}^{0}_{-1}e$

15. (a) $^{0}_{-1}e$, (b) $^{4}_{2}He$, (c)   , (d) $^{0}_{-1}e$

16. Weighing techniques are not sensitive enough to detect small mass losses.

17. Compared to our short lifetimes, history, and the projected future of the solar system, it is stable.

18. The short-lived isotopes are products of decay series. The original isotopes presumably did decay.

19. 1 Ci = 3.7 x 10^{10} decays/s = 3.7 x 10^{10} Bq.

20. 25,000 - 30,000 years is $5t_{1/2}$ - $6t_{1/2}$ for C-14.

    After this many half-lives, the activity is very small and is difficult to get accurate data.

21. There would be more C-14 than thought and a greater activity (or apparently fewer half-lives) when measured, so the ages of objects would be computed to be less than they are.

22. Strong nuclear force: pi muon (or pion), weak nuclear forces: W-particle, electromagnetic force: special photon, and the gravitational force: graviton.

23. Quarks. No, they have fractional electronic charges. For example, the quarks that make up the proton are believed to carry charges of + 2/3 and - 1/3.

24. Six flavors; up (u), down (d), strange (s), charmed (c), beauty or bottom (b), and truth or top (t). For each flavor, there are three colors or editions, so 6 x 3 = 18, and with each quark having an antiparticle antiquark, 2 x 18 = 36.

SAMPLE TEST QUESTIONS

Multiple Choice

1. Rutherford investigated the structure of the atom by means of (a) X-rays, (b) the photoelectric effect, *(c) particle scattering, (d) ultraviolet light.

2. The nuclear force acts only between (a) protons, (b) neutrons, (c) protons and neutrons, *(d) all of the preceding.

3. Which of the following nuclear radiations has the largest electrical charge? *(a) alpha particle, (b) beta particle, (c) gamma particle, (d) neutron.

4. The term nucleon applies to (a) electrons, (b) only protons, (c) only neutrons, *(d) both nuclear protons and neutrons.

5. A type of radioactive decay in which the mass numbers of the parent and daughter nuclei are different is *(a) alpha decay, (b) beta decay, (c) gamma decay.

6. A type of radioactive decay in which both the proton and neutron numbers of the parent and daughter nuclei are different, but the mass numbers are the same is (a) alpha decay, *(b) beta decay, (c) gamma decay.

7. Which of the following does not remain constant in every nuclear decay process? (a) total number of nucleons, (b) total charge, *(c) total number of neutrons, (d) total mass energy.

8. At the end of one half-life, a sample of radioactive material (a) is no longer radioactive, (b) is half as massive, (c) will have a shorter half-life, *(d) has one-half of its initial activity.

9. A type of radiation detector that uses visible flashes of light is a (a) Geiger counter, *(b) scintillation counter, (c) bubble chamber, (d) solid state detector.

10. The theoretical exchange particle in the gravitational interaction is the (a) W-particle, (b) pion, (c) gluon, *(d) graviton

11. Radioactivity was discovered by (a) Rutherford, *(b) Becquerel, (c) Curie, (d) Geiger.

12. The idea of the atomic nucleus was investigated experimentally by *(a) Rutherford, (b) Becquerel, (c) Geiger, (d) Gell-Mann.

13. Isotopes of an element have the same (a) number of nucleons, (b) number of neutrons, *(c) number of protons, (d) mass.

14. The nuclear force acts between (a) proton-proton pairs, (b) proton-neutron pairs, (c) neutron-neutron pairs, *(d) all of the preceding.

15. Which of the following is not conserved in all nuclear reactions? (a) charge, (b) mass-energy, *(c) neutrons, (d) nucleons.

16. The most massive decay particle is the *(a) alpha particle, (b) beta particle, (c) gamma particle, (d) pion.

17. The relatively weakest fundamental interaction is the (a) strong nuclear, *(b) gravitational, (c) electromagnetic, (d) weak nuclear.

18. In three half-lives, the radioactivity of a sample decreases by a factor of (a) 1/2, (b) 3/4, (c) 5/6, *(d) 7/8.

61-5

19. A Geiger counter counts by means of (a) vapor bubbles, (b) flashes of light, *(c) current pulses, (d) electron-hole pairs.

20. The exchange particle for the strong nuclear interaction is the (a) quark, *(b) pion, (c) neutrino, (d) gluon.

21. The central core of the atom (a) is composed of protons and neutrons, (b) possesses a positive charge, (c) has no electrons, *(d) all of these.

22. The proton number of an atom (a) is equal to the neutron number, (b) is always greater than the neutron number, (c) is always less than the neutron number, *(d) defines its atoms as being a particular element.

23. The mass number is the sum of the (a) protons and electrons, (b) neutrons and electrons, *(c) protons and neutrons, (d) protons, neutrons, and electrons.

24. An alpha particle (a) has an atomic mass of four, (b) carries a plus two electronic charge, (c) is composed of two protons and two neutrons, *(d) all of these.

25. A beta particle (a) carries a positive electronic charge, *(b) is an electron, (c) is more massive than the alpha particle, (d) is a proton.

26. A gamma "particle" (a) is not a particle, (b) carries no electronic charge, (c) is a photon of energy, *(d) all of these.

27. Which of the following is not a fundamental force (a) electromagnetic, (b) gravitational, *(c) radioactive, (d) strong (nuclear), (e) weak (nuclear).

28. The elementary particle or particles responsibility for fundamental forces or interactions is/are the (a) gluon, (b) photon, (c) graviton, (d) W and Z, *(e) all of these.

29. Quarks are believed to (a) carry no electronic charge, (b) have a mass equal to that of the electron, *(c) be the fundamental particles that make up all hadrons, (d) none of these.

30. The exchange particles for the weak force are *(a) W and Z, (b) photon and graviton, (c) gluon and graviton, (d) U and D quarks.

31. The Grand Unified Theory (GUT) is a theory that unifies the (a) electromagnetic force and the gravitational force, *(b) electroweak force and the strong gluon force, (c) electromagnetic force and the electroweak force, (d) all of these.

Completion

1. $^{238}_{92}U$ has <u>92</u> protons and <u>146</u> neutrons in its nucleus.

2. The Z number is equal to the number of <u>protons</u>.

3. Uranium-238 and uranium-235 are <u>isotopes</u> of uranium.

4. The half-life of a radioactive isotope is 1 hour. The fraction of the isotope left in a sample after one hour is <u>one-half</u> the original amount; and after two hours, <u>one-fourth</u> the oroginal amount; and after three hours, <u>one-eight</u> the original.

5. The force between pairs of nucleons is the <u>strong nuclear</u> force.

6. The Curies discovered the radioactive elements <u>radium</u> and <u>polonium</u>.

7. The time required for a detector to recover for another detection or count is called the <u>dead time</u>.

8. The sum of the number of protons and neutrons in a nucleus is equal to the <u>mass</u> number of the nucleus.

9. Nuclei which spontaneously decay with the emission of energetic particles are said to be <u>radioactive</u>.

10. One becquerel is equal to <u>one decay per second</u>.

11. A nuclide species is determined by the <u>proton number</u>.

12. Tritium is a(n) <u>isotope</u> of hydrogen.

13. An alpha particle has a <u>positive, +2</u> electric charge.

14. In beta decay, the neutron number <u>decreases by one</u>.

15. The strong nuclear force has a <u>short</u> range.

16. Plants and animals contain radioactive <u>carbon (14)</u>.

17. A Geiger counter detects radiation by <u>ionization</u> in a gas.

18. Radioactive <u>iodine (131)</u> is used to monitor thyroid functions.

19. The exchange particle for the strong interaction is the <u>pion (pi muon)</u>.

20. Isotopes of an element have the same number of <u>protons</u>.

21. The nuclear force acts between <u>protons</u> and <u>neutrons</u>.

22. An alpha particle is the nucleus of a <u>helium</u> atom.

23. The total number of nucleons remains <u>constant</u> in a nuclear reaction.

24. The radioactivity of a material decreases by <u>3/4</u> in two half-life.

25. The becquerel is a unit of <u>radioactivity</u>.

26. Quarks have charges of <u>-1/3 and +2/3</u>.

27. The number of protons in $_1^3$H is <u>equal to</u> the number of protons in $_1^2$H.

28. Isotopes of an element have different numbers of <u>neutrons</u>.

29. The neutron number is conserved in <u>alpha or gamma</u> decay.

30. A proton is through to be made up of three charged <u>quarks</u>.

31. Exchange <u>particles</u> are responsible for the four fundamental interactions.

32. The daughter nucleus of carbon-14 decay is <u>nitrogen-14</u>.

33. The exchange particles for the weak force are the
    <u>W and Z particles</u>.

34. The exchange particle for the gravitational force
    is called the <u>graviton</u>.

35. The theory which unifies the electroweak force with
    the strong gluon force is known as the
    <u>grandunified theory</u>.

36. The number of protons plus the number of neutrons
    in an atoms is known as the <u>mass number</u>.

37. An alpha particle consists of <u>two protons and two
    neutrons</u>.

38. A gamma "particle" is a quantum of <u>electromagnetic
    energy</u>.

39. The decay rate of a radioactive isotope is measured
    in terms of a characteristic time called the
    <u>half-life</u>.

40. All strongly interacting particles, which include
    the nuclear protons and neutrons, are called
    <u>hadrons</u>.

41. The theoretical exchange particle between quarks is
    the <u>gluon</u>.

Chapter 25

Matching

(Choose the appropriate answer from the list on the right.)

<u>a</u> 1. type of decay with most massive particle

<u>c</u> 2. no change in proton number

<u>a</u> 3. decay particle with *+2* positive charge

<u>c</u> 4. has no electric charge

<u>c</u> 5. most penetrating

<u>b</u> 6. has proton *charge* number of -1

<u>b</u> 7. $^{234}_{90}\text{Th} \;\text{--->}\; ^{234}_{91}\text{Pa} + \underline{\hspace{1cm}}$

<u>a</u> 8. $^{238}_{92}\text{U} \;\text{--->}\; ^{234}_{90}\text{Th} + \underline{\hspace{1cm}}$

<u>c</u> 9. a quantum of electromagnetic energy

<u>a</u> 10. a helium nucleus

a. alpha (particle or decay)

b. beta (particle or decay)

c. gamma (particle or decay)

d. positron

238

Matching

(Choose the appropriate answer from the list on the right.)

_n_	1.	Rutherford
_d_	2.	neutron
_k_	3.	isotope
_a_	4.	deuterium
_i_	5.	nuclear force
_b_	6.	Becquerel
_j_	7.	alpha particle
_e_	8.	beta particle
_c_	9.	gamma particle
_f_	10.	quark

a. hydrogen

b. radioactivity

c. quantum

d. nucleon

e. electron

f. fractional electric charge

g. Geiger counter

h. proton

i. short range

j. helium nucleus

k. same proton number

l. carbon-12

m. half-life

n. nucleus

o. neutrino

Chapter 25

Matching

(Choose the appropriate answer from the list on the right.)

d   1. hadrons

h   2. gluon

e   3. W and Z particles

b   4. photon

i   5. graviton

a   6. electroweak force

g   7. grandunified theory (GUT)

j   8. superforce

f   9. exchange particles

c   10. u and d particles

a. incorperates the electro-magnetic and weak forces

b. interacts for the electro-magnetic force

c. quarks

d. all are strongly interacting particles

e. interacts for the weak force

f. responsible for fundamental forces or interactions

g. combines the strong electro-weak forces

h. interacts for the strong force

i. interacts for the gravitation force

j. describes all fundamental interactions

Answers to Exercises

1. (a) 50 µg, (b) 25 µg, (c) 12.5 µg.

2. $t_{\frac{1}{2}}$ = 5.3 y for Co-60 (Table 25.2). Then 120 --->
    60 ---> 30 ---> 15 µCi, and 15 µCi remains at the
    end of $3t_{\frac{1}{2}}$ (as indicated by arrows), and $3t_{\frac{1}{2}}$ =
    3(5.3) = 15.9 y.

3. (a) In 4s, from 24 ---> 12 ---> = 6 bubbles, or
    $2t_{\frac{1}{2}}$, and $t_{\frac{1}{2}}$ = 2s. (b) $3t_{\frac{1}{2}}$ and 3 bubbles.

4. In 12 hours, 1600 ---> 800 ---> 400 ---> 200 cps,
    or $3t_{\frac{1}{2}}$ have elapsed (as indicated by arrows), so
    3 $t_{\frac{1}{2}}$ = 12 h or $t_{\frac{1}{2}}$ = 4 h.

5. With $t_{\frac{1}{2}}$ = 10 min, 1 hour = 60 min = $6t_{\frac{1}{2}}$. Working
    in terms or fractions and taking the activity to be
    1.0 (100%) at t = 0, then 1 ---> 1/2 ---> 1/4 --->
    1/8 ---> 1/16 ---> 1/32 ---> 1/64 ---> and 1/64
    (0.016 or 1.6%) of the initial activity at $6t_{\frac{1}{2}}$ (as
    indicated by arrows). Hence the activity
    decreases by 98.4%.

6. $t_{\frac{1}{2}}$ = 8 days for I-131, so in one month (30 or 31
    days), the sample undergoes slightly less than $4t_{\frac{1}{2}}$.
    Then, 100 ---> 50 ---> 25 ---> 12.5 ---> 6.3 µg
    remain at the end of $4t_{\frac{1}{2}}$ (as indicated by arrows),
    so slightly less than this amount remains at the
    end of one month.

7. With 16 beta emissions/g of C/min initially,

    16 ---> 8 ---> 4 ---> 2, and the C-14 has gone

    through $3t_{\frac{1}{2}}$ (as indicated by arrows) since the tree

    was fallen. With $t_{\frac{1}{2}}$ = 5730 y for C-14 (Table

    28.2), $3t_{\frac{1}{2}}$ = 3(5730) = 17,190 y, so the carving is

    on the order of 17,000 years old.

8. With 16 beta emissions/g of C/min when the animal
    was alive, 16 ---> 8 ---> 4, or $2t_{\frac{1}{2}}$ would elapse

    before the beta decay rate was 4 beta emissions/g
    of C/min. With $t_{\frac{1}{2}}$ = 5730 y for C-14 (Table 25.2),

    $2t_{\frac{1}{2}}$ = 2(5730) = 11,460 y, and taking the current

    year to be 1990, we have the year 1, 990 + 11,460 =
    13,450 A.D.

Answers to Questions

1.   The pros focus on general freedoms -- speech,
     writing, etc., and the cons focus on defense and
     national security, (and perhaps ethical issues) in
     the nuclear case. It will be interesting to see
     what your students consider to be pros and cons.

2.   (a) Hahn -- verified the fission reaction, (b)
     Meitner -- worked out the theory of nuclear
     fission, (c) Fermi -- studied neutron bombardment
     and was a key figure in the Manhattan Project, (d)
     Bohr -- transmitted the discovery or theory of
     nuclear fission to scientists in America, (e)
     Einstein -- wrote a letter to President Roosevelt
     describing the potential of nuclear fission.

3.   No, in large part in Illinois at the University of
     Chicago.

4.   A controversial question that stimulates various
     student opinions, particularly since most
     present-day college students did not experience
     WW-II.

5.   As evidenced in the Hint, it would not be a good
     buy for long-term security.

6.   This would depend on the cross-section of the
     reaction with regard to "fast" and "slow"
     particles. In terms of force and energy, if an
     alpha particle and a proton were accelerated by the
     same electric field through equal distances, since
     $E = F/q$ or $F = qE$, the alpha particle would
     experience twice the force and hence have twice the
     energy. However, since an alpha particle is about
     four times as massive as a proton, the proton would
     have a greater speed (by a factor of 1.4 or

     $\sqrt{2}$). $Fd = K = \frac{1}{2}mv_p^2$ for the proton, and $2Fd = K =$

     $\frac{1}{2}(4m)v_a^2$ and $v_a^2 = 2\ v_a^2$.

7.   Conservation of nucleons means that the total mass
     numbers on each side of a nuclear equation are the
     same. Conservation of charge means that the total
     proton numbers on each side of a nculear equation
     are the same. (The neutron number is not always
     conserved, e.g., in beta decay.)

8. (a) $_{22}^{48}\text{Ti}$, (b) , (c) $_{16}^{32}\text{S}$, (d) $_{56}^{141}\text{Ba}$,

(e) $_{2}^{3}\text{He}$

9. $_{0}^{1}\text{n} + _{92}^{235}\text{U} \longrightarrow _{37}^{94}\text{Rb} + _{55}^{139}\text{Cs} + 3(_{0}^{1}\text{n})$

10. They ran out of known "trans-Uranus" planets.

11. These should be evident from this and previous chapters.

12. Because Ru stands for ruthenium.

13. Relatively very small, 1 eV = 1.6×10^{-19} J.

14. A series of branching, connecting matches similar to the diagram of a nuclear chain reaction.

15. Not a critical mass and neutrons escape.

16. Uranium enrichment and confinement are not adequate to allow a build up for explosion. (See fission Reactors section in text.)

17. Yes, by a controlling moderation and the number of "slow" neutrons, the fissioning of U235 nuclei and the rate of the chair reaction is affected.

18. In heavy water (D_2O) reactors, natural (unenriched)

uranium can be used. However, in light water (ordinary H_2O) reactors, enriched uranium is

required, since the hydrogen nuclei of ordinary water have a tendency to capture neutrons. (See Fission Reactors section in text.)

19. An uncontrolled reactor condition in which the fuel rods fuse and the fissioning mass melts through the reactor floor into the environment. The "China Syndrome" is used to emphasize a meltdown situation with the false implication that the fissioning mass might melt through the Earth to China. (This would not occur, not only because China is not opposite the U.S. through the center of the Earth.)

20. Because of a LOCA and there may be sufficient heat generated to fuse the fuel rods.

21. Various pros, e.g., less dependence on foreign oil, less air pollution than from coal-fired plants, less refueling, etc. Various cons, e.g., possible meltdown and pollution, radioactive waste disposal, nuclear proliferation, etc.

22. To provide the required energy for the high temperatures to initiate fusion.

23. $_{0}^{1}n + _{3}^{6}Li --> _{1}^{3}H + _{2}^{4}He$ (See Fusion and Fusion Reactors section in text.)

24. In general there is more energy from fusion than from fission in a "kilogram per kilogram" comparison (many more nuclei in a kilogram of fusion material than in a kilogram of fission material, since nuclei are much heavier), and nuclear wastes from fusion are less a problem (shorter half-lives) than from fission.

25. A fusion reactor would be quickly shut down by restricting fuel input and heat would be quickly dissipated from the high temperature plasma region that would curtail the fusion reaction. Also, there is no possibility of a meltdown.

26. Magnetic confinement refers to the use of a magnetic field to hold a hot plasma in a confined space. Inertial confinement refers to the inertia given to fuel pellets by means of a laser, electron, or iron beam.

27. By the gravity of the dense mass.

28. Gravity collapses the stellar gases and increases the temperature until fusion in initiated.

29. Ar-90 is chemically similar to Ca-40 and bones have a high calcium content.

30. Iodine, when taken into the body is readily concentrated in the thyroid gland. Thus, the thyroid gland would be over exposed to the radiation from the I-131.

31. Natural background radiation comes from radioactive elements in the Earth, cosmic rays, and atmosphere reactions of solar radiations. "Unnatural" radiation comes from nuclear explosions or form a nuclear reactor accident.

Chapter 26

SAMPLE TEST QUESTIONS

Multiple Choice

1. The first experimental artificial transmutation of a nucleus was done by (a) Fermi, *(b) Rutherford, (c) Einstein, (d) Curie.

2. The first induced nuclear reaction was initiated by (a) particles from an accelerator, (b) high-temperature, *(c) radioactive decay particles, (d) chemical means.

3. Which of the following is a fissionable isotope? (a) C-12, (b) deuterium, (c) Au-179, *(d) Pu-239.

4. In a nuclear reactor, the neutrons are slowed down by the (a) chain reaction, (b) control rods, *(c) moderator, (d) exoergic reaction.

5. The chain reaction in a nuclear reactor is controlled by controlling the *(a) number of available neutrons, (b) critical mass, (c) amount of coolant, (d) meltdown.

6. To have a sustained fission reaction, there must be (a) beta decay, (b) control rods, *(c) a critical mass, (d) more U-235 than U-238.

7. The principle of the hydrogen bomb is (a) radioactive decay, (b) fission, *(c) fusion, (d) chemical explosion.

8. A nuclide commonly associated with nuclear proliferation is (a) U-238, (b) U-235, *(c) Pu-239, (d) tritum.

9. A problem with energy production from nuclear fission that would not be so severe with nuclear fusion is (a) nuclear weapons, *(b) longevity of nuclear wastes, (c) nuclear test ban treaties, (d) nuclear proliferation.

10. A major problem in the development of a fusion reactor is *(a) confinement, (b) fuel availability, (c) nuclear wastes, (d) D-T reactions.

11. The person who worked out the theory of nuclear fission was (a) Rutherford, (b) Fermi, (c) Einstein, *(d) Meitner.

12. The first fission reactions were produced by (a) Bohr, *(b) Fermi, (c) Rutherford, (d) Meitner.

13. The missing product in the reaction

$$_0^1n + _{13}^{27}Al \longrightarrow _{12}^{27}Mg + \underline{\hspace{2cm}}$$

65-3

is (a) $_{-1}^0e$, (b) $_0^1n$, *(c) $_1^1H$, (d) $_2^4He$.

14. Which of the following is not associated with a fission reactor? *(a) plasma, (b) critical mass, (c) control rods, (d) moderator.

15. The principle of the hydrogen bomb is (a) fission, *(b) fusion, (c) magnetic confinement, (d) fertile material.

16. A fertile material is (a) Pu-239, (b) U-233, (c) U-235, *(d) U-238.

17. A serious problem with a nuclear fission reactor could be (a) the possibility of exploding like a bomb, (b) heavy water, *(c) a LOCA, (d) deuterium.

66-2

18. A hydrogen fusion reaction does not require *(a) a critical mass, (b) a plasma, (c) a high temperature, (d) deuterium.

19. The "splitting" of the atom refers to (a) nuclear waste, (b) endoergic reactions, (c) fusion, *(d) fission.

63-2

4

20. The moderator of a reactor is used to (a) supply critical mass, *(b) produce slow neutrons, (c) initiate a chain reaction, (d) remove radioactive waste.

21. Which of the following is <u>not</u> a unit of energy? (a) joule, (b) electron volt, *(c) watt, (d) calorie.

22. A nuclear reaction in which there is a net energy input is known as a _____ reaction. (a) fission, (b) fusion, (c) exoergic, *(d) endoergic.

23. A nuclear reaction in which ther is a net energy output is known as a _____ reaction. (a) fusion, (b) fission, *(c) exoergic, (d) endoergic.

24. The energy needed to separate a nucleus into free protons and neutrons is called _____ energy. (a) potential, *(b) binding, (c) nuclear, (d) internal.

61-2

8

25. The amount of mass or concentration of fissionable material needed for a sustained chain reaction is known as the _____ mass. (a) fissionable, (b) reaction, *(c) critical, (d) nuclear.

26. A reactor that produces more fuel than it consumes by converting nonfissionable nuclei into fissionable nuclei is called a _____ reactor. (a) atomic, (b) fission, *(c) breeder, (d) fusion.

27. A nuclear reactor cannot explode like an atomic or nuclear bomb. This is due to the fact that reactor-grade uranium contains only about _____ percent uranium 235. *(a) 3, (b) 8, (c) 12, (d) 15.

28. The most likely fuel to be used for a fusion reaction is *(a) hydrogen, (b) helium, (c) uranium, (d) plutonium.

29. Magnetic confinement refers to (a) the density of a plasma in a magnetic field, *(b) a plasma held in a confined space of a magnetic field, (c) plasma temperature in a magnetic field, (d) a technique used to initiate fusion.

30. Which of the following is not a fissionable fuel? (a) uranium-235, (b) plutonium-239, (c) uranium-233, *(d) all of the above are fissionable fuels.

Completion

1. The elements with proton numbers greater than 92 are referred to as <u>transuranic</u> elements.

2. $^{235}_{92}U + ^{1}_{0}n --> ^{141}_{56}Ba + ^{92}_{36}Kr + 3 ^{1}_{0}n$

3. $2 ^{3}_{1}H --> ^{4}_{2}He + 2 ^{1}_{0}n$

4. In a fission reaction, mass is converted into <u>energy</u>.

5. An example of uncontrolled nuclear fission is a <u>nuclear explosion (bomb)</u>.

6. A major problem in the development of fusion reactors is <u>confinement</u>.

7. Spent fuel rods are an example of <u>nuclear waste</u>.

8. D_2O, commonly known as <u>heavy water</u>, can be used as a moderator in a reactor using <u>natural</u> uranium.

9. An <u>endoergic</u> nuclear reaction is one in which <u>energy</u> is converted to <u>mass</u>.

10. The potential method of initiating nuclear fusion using laser or electrical beams is called <u>inertial confinement</u>.

11. The first induced nuclear reaction was performed by <u>Lord Raleigh</u>.

12. A common unit of energy used for accelerated particles is the <u>electron volt or joule</u>.

13. Energy must be put into a(n) <u>endoergic</u> reaction.

14. Sustained energy release by fission results from a <u>chain</u> reaction.

15. Nuclear reactor energy output is controlled by means of <u>control rods</u>.

16. Heavy water is used as a <u>moderator</u> in some nuclear reactions.

17. Fissionable material is produced in a breeded reactor from <u>fertile</u> material.

18. Spent fuel rods from reactors come under the classification of <u>nuclear waste</u>.

19. A plasma is a gas of <u>charged particles</u>.

20. A LOCA could lead to a <u>meltdown</u>.

21. The electron volt is a unit of <u>energy</u>.

22. Fission is an <u>exoergic</u> reaction.

23. A chain reaction requires a <u>critical</u> mass.

24. The 1963 Nuclear Test Ban Treaty banned <u>atmospheric</u> nuclear testing.

25. The symbol $_{+1}^{0}e$ represents a <u>positron</u>.

26. Fusion reactions are called <u>thermonuclear</u> reactions.

27. <u>Magnetic</u> confinement for fusion uses magnetic fields.

28. <u>Cross-section</u> is a measure of the probability for a reaction to occur.

29. A gas consisting almost entirely of positively charged ions and free, negatively charged electrons, is called <u>plasma</u>.

30. A fusion weapon is commonly called a <u>thermonuclear</u> device.

31. The fuel for a common nuclear reactor is <u>uranium-235</u>.

32. A breeder reactor converts U-238 into <u>Pu-239</u> which is fissionable.

33. The biological hazard of nuclear radiation depends on the radiation energy, its electrical charge, and <u>density</u> of the material.

34. A critical mass is required for a <u>fission</u> reaction.

35. The source of stellar energy is <u>fusion</u>.

36. A lot of energy is needed to fuse nuclei togetehr because the <u>repulsive force</u> existing between the nuclei must be overcome.

37. A LOCA can give rise to a reactor <u>meltdown</u>.

38. The energy an electron receives when accelerated through a potential of one volt is called <u>one electron volt</u>.

39. The energy needed to separate a nucleus into free photons and neutrons is called <u>binding</u> energy.

40. The changing of the nuclei of one element into the nuclei of another element through an induced nuclear a reaction is called <u>artificial</u> transmutation.

Matching

(Choose the appropriate answer from the list on the right.)

_e_ 1. a fission reaction   a. boron

_j_ 2. increased availability of radioactive material   b. breeder reactor

         c. critical mass

_g_ 3. the combining of two light nuclei to form a heavier nucleus   d. endoergic

         e. exoergic

_b_ 4. generates more fuel than consumed   f. fission

_h_ 5. a moderator   g. fusion

_a_ 6. used in constructing control rods   h. graphite

         i. binding energy

_c_ 7. sufficient fissionable material for a chain reaction   j. nuclear proliferation

_i_ 8. energy needed to separate a nucleus into free protons and neutrons   k. nuclear reactor

         l. plasma

_l_ 9. a gas of charged particles   m. Manhattan project

         n. uranium

_m_ 10. atomic bomb development   o. nuclear waste

Chapter 26

Matching

(Choose the appropriate answer from the list on the right.)

_h_ 1.	atomic bomb	a. controlled fission
_a_ 2.	nuclear reactor	b. slow neutrons
_e_ 3.	Fermi	c. electron volt
_l_ 4.	Rutherford	d. plutonium-239
_j_ 5.	exoergic	e. artificial isotopes
_b_ 6.	moderator	f. anti-nuke
_k_ 7.	D_2O	g. LOCA
_g_ 8.	meltdown	h. uncontrolled fission
_d_ 9.	breeder reactor	i. leukemia
_m_ 10.	fusion	j. energy out
		k. heavy water
		l. first induced reaction
		m. plasma
		n. radioactive waste
		o. energy in

Chapter 27   Astrophysics

Answers to Questions

1. The source of cosmic rays are unknown (most likely they come from supernovae). They are "rays" (largely protons) that come to us from the universe (cosmos), hence the name cosmic rays.

2. The known or observable universe prior to the intervention of the telescope was limited to observations made with the unaided eye. Quasars set the present limit of 16 billion light years (if the distance as computed from the Doppler shift and Hubble's law is correct).

3. Yes, since the light or information we currently receive is "old" or as old as the time it takes to reach us. If a star 100 l.y. away "died", we would not know it for 100 years.

4. Not a horizontal straight line would indicate no change in or a constant value of H, but data shows that Hubble's "constant" is smaller today than it was in the past due to the expansion of the universe.

5. We cannot really tell where the center is since observations at different locations in the expanding universe appear similar.

6. No, the Big Bang may have been the creation of space itself.

7. (a) You will observe the other two cars receeding from you in opposite directions. (b) The observer in the car behind you will observe the two cars ahead to be receding in the same direction, and the car in front will see you and the car behind receding in the same, but opposite direction.

8. (a) The average density decreases with the Big Bang theory since the universe is expanding. (b) It remains steady in the steady state theory as a result of new matter being created. (c) The average density would "oscillate" (get smaller on expansion and larger on contraction) in the oscillating universe theory.

9. Yes, in the steady state theory, mass is not conserved with new matter being "created" out of nothing.

10. In an oscillating universe, Hubble's constant would also "oscillate", i.e., get smaller on expansion and larger on contraction.

11. The parallel rails appear to converge and meet in the distance as would parallel light beams in a closed universe.

12. The steady state theory predicts an ever-expanding universe.

13. A protogalaxy becomes a galaxy when a sufficient number of stars are formed. A protostar becomes a star when fusion is initiated.

14. It takes a long time for the gases to gravitationally callapse to the point that fusion can be initiated.

15. It doesn't really. Stellar evolution is allowed by both theories.

16. The Sun will become a white dwarf, since at this stage it will have less mass than the 1.4 solar masses needed for further fusion.

17. Primarily through the radiation of energy resulting from the fusion process in which mass is converted into energy.

18. If the Earth remained at its 93 million mile distance from the Sun, it would be engulfed in the outer hydrogen layer of the Sun.

19. The fate of a star depends on its mass. Relatively small stars become white dwarfs. Larger or more massive stars supernova with the formation of neutron stars and possibly black holes.

20. In the last 15 billion years, matter has spent a great deal of time collapsing to form galaxies and stars. In the future it will be recycled through stellar evolution and become less dense on the average due to expansion.

21. In the sense that overweight persons may have shorter life times as do very massive stars.

22. The gravity would be the same at the distance of the radius of the original star, but the shrinking into a black hold allows closer approach toward the center and $F \propto 1/r^2$ so the force gets greater.

23. Not really in the sense that we do not know if Hubble's law holds for very large distances. As suggested in the hint, by analogy Hooke's law, F = kx, only applies up to the elastic limit of the spring and it would not be justified to extrapolate a F versus x plot beyond this.

24. Quasar (quasi-stellar radio source) -- quasar is considered to be an extragalactic, highly luminous object that exhibits a large red-shift.

    Pulsar -- a rapidly rotating star that emits periodic pulses of electromagnetic radiation.

    Neutron star -- the remains of a supernova explosion..... an extremely dense mass of almost pure neutrons.

    Black hole -- a mass that has collapsed, due to its own gravitational force, to a small volume such that light or any electromagnetic radiation cannot escape from its surface.

25. For a light day, $d = ct = (3 \times 10^5$ km/s)(24 h) $(3.6 \times 10^3$s/h) $= 2.6 \times 10^{10}$ km, and 2.6×10^{10} km $(1$ mi/1.6 km) $= 1.6 \times 10^{10}$ m. A light week is then seven times a light day, so $7(2.6 \times 10^{10}) = 1.8 \times 10^{11}$ km and $7(1.6 \times 10^{10}) = 1.1 \times 10^{11}$ mi.

    One light second is 3×10^5 km, so the diameter of the Sun is on the order of:

    10^6 km $(1$ light s/3 $\times 10^5$ km) $= 3.3$ light second.

26. Detect effects of the black hole's gravity such as the distortion of light coming from background stars. Also, in a double star system, a black hole capturing material from its binary neighbor produces X-rays, which could be observed.

SAMPLE TEST QUESTIONS

Multiple Choice

1. Astrophysics is used in the explanations of (a) cosmology, (b) the expanding universe, (c) the cosmos, *(d) all of the preceding.

2.   Hubble's constant would be expected to be unchanging with time in the (a) Big Bang theory, (b) Steady State theory, (c) Oscillating theory, *(d) all of the preceding.

3.   The maximum theoretical age of the universe is about (a) 100 million years, (b) 5 billion years, *(c) 12 billion years, (d) 50 billion years.

4.   The age of the universe is estimated from (a) the Steady State theory, (b) pulsars, (c) 3-K radiation, *(d) Hubble's constant.

5.   The critical factor in the Oscillating Universe theory is (a) Hubble's constant, *(b) mass, (c) hydrogen fusion, (d) temperature.

6.   A big bang, expanding universe, and big crunch are all parts of the (a) Big Bang theory, (b) Steady State theory, *(c) Oscillating theory, (d) stellar evolution.

7.   Pockets of contracted gas that do not have enough mass for the contraction to produce the temperature for fusion ignition are called *(a) brown dwarfs, (b) black dwarfs, (c) white dwarfs, (d) novae.

8.   Relatively small stars eventually become (a) pulsars, (b) supernovae, (c) quasars, *(d) black dwarfs.

9.   Heavier elements above iron are formed by *(a) supernovae, (b) pulsars, (c) white dwarfs, (d) neutron stars.

10. The most distant objects we can observe in the universe are believed to be *(a) quasars, (b) pulsars, (c) neutron stars, (d) black holes.

11. An expanding universe is supported by (a) supernovae, *(b) Doppler shifts, (c) the event horizon, (d) the carbon flash.

12. Hubble's constant has units of *(a) s^{-1}, (b) m/s, (c) m, (d) m/s^2

13. A critical average density is necessary in the (a) Big Bang theory, (b) steady-state theory, *(c) oscillating universe theory, (d) all of the preceding.

14. Our Sun will eventually become a (a) supernova, (b) brown dwarf, (c) pulsar, *(d) red giant.

15. Which of the following phases of stellar evolution comes latest? (a) red giant, (b) hydrogen burning, *(c) black dwarf, (d) white dwarf.

16. The supernova stage of a star, (a) occurs after becoming a white dwarf, (b) will occur for our Sun, (c) is a quasar, *(d) does not occur for small stars.

17. Quasars are (a) remanents of supernovae, *(b) radio galaxies, (c) black holes, (d) distant red giants.

18. Pulsars are believed to be (a) red giants, (b) quasars, *(c) neutron stars, (d) the origin of 3-K radiation.

19. A black hole does not have (a) a low density, *(b) an event cone, (c) an accretion disk, (d) an event horizon.

20. If Hubble's law is applicable, the most distant, observable object is a *(a) quasar, (b) pulsar, (c) supernova, (d) black hole.

21. Evidence to support the Big Bang theory of the universe is the (a) cosmological red-shift, (b) 3-K cosmic microwave background radiation, (c) hydrogen to helium mass ratio of three to one, *(d) all of these.

22. Hubble's law states that the (a) recessional velocity of a galaxy times its distance is a constant, *(b) ratio of the recessional velocity of a galaxy to its distance is a constant, (c) distance to a galaxy divided by the recessional velocity of the galaxy is a constant, (d) none of the above.

23. The event horizon *(a) refers to the outer boundary of the black hole, (b) is equal to the radius of the black hole, (c) is the same as the event cone, (d) all of these.

24. When observation of the universe is made over a very large scale, the distribution of matter (galaxies) is (a) isotropic, (b) homogeneous, *(c) both isotropic and homogeneous, (d) none of the above.

25. The hydrogen "burning" stage of a star, such as the Sun, accounts for about ___ percent of its lifetime. (a) 25, (b) 50, (c) 75, *(d) 90

26. A supernova (a) is an exploding star, (b) is the most energetic of all stellar explosions, (c) occurs for a star that is a few times more massive than the sun, *(d) all of these.

27. A neutron star (a) is composed of almost pure neutrons, (b) is believed to be the remains of a supernova, (c) has a very high density, *(d) all of these.

28. The steady-state theory of the universe (a) views the universe as having no beginning or end, (b) states that the universe is the same everywhere, (c) states that the universe is same at all times, *(d) all of these.

29. The Schwarzchild radius refers to (a) the radius of the universe, (b) the radius of a star, (c) the critical size for the formation of star, *(d) the critical size at which a shrinking star becomes a black hole.

Completion

1. Cosmology is the study of the nature and structure of the universe.

2. The fuel used in the energy production for most stars is hydrogen.

3. The expanding universe is evidenced by the red shift of light from galaxies.

4. The age of the universe is computed using Hubble's constant which gives the age to be 12 billion years.

5. The life time of a massive star is generally less than that of a less massive star.

6. When a star exhausts its supply of hydrogen, it becomes a red giant.

7. Neutron stars are believed to result from super-novae explosions.

8. The most distant objects in the universe according to the Hubble relationship are quasars.

9. The Big Bang theory predicts a(n) open universe and the oscillating theory predicts a(n) closed universe.

10. In the final stage of a dying star, it will either become a <u>black dwarf</u> or a <u>supernova</u>.

11. The Big Bang theory was originated by <u>Lemaitre</u>.

12. Doppler red shifts indicate a(n) <u>expanding universe</u>.

13. The Hubble constant puts the age of the universe to be about <u>10-12 billion</u> years.

14. The oscillating universe theory requires the universe to have a critical <u>(average) density</u> for expansion reversal.

15. Stellar birth occurs with the initiation of <u>fusion</u>.

16. A star with greater than 1.4 solar masses will eventually explode as a <u>supernova</u>.

17. A white dwarf eventually becomes a <u>black dwarf</u>.

18. Pulsars are believed to be <u>nuetron</u>.

19. Star-like radio sources are called <u>quasars</u>.

20. The boundary of a black hole is the surface of its <u>event horizon</u>.

21. The critical temperature at which a star is "born" is the temperature at which <u>fusion</u> begins.

22. The observation that galaxies are moving away from Earth at high speeds supports the <u>expanding</u> universe theory.

23. In older stars, fusion takes place between helium nuclei producing <u>carbon</u> nuclei.

24. Hubble's law states that the speed of a galaxy a distance d away has <u>twice</u> the speed of a galaxy a distance d/2 away.

25. According to the <u>Steady State</u> theory, new matter is steadily created.

26. The discovery of cosmic background radiation supported the <u>Big Bang</u> theory.

27. In an open universe, two parallel light beams would <u>diverge</u>.

28. The Crab Nebula is believed to have been the result of a <u>supernova</u>.

29. The boundary of a black hole is the surface of its event horizon which is at the <u>Schwarchild</u> radius.

30. The Sun will eventually become a <u>black dwarf</u>.

31. The study of the nature and structure of the universe is called <u>cosmology</u>.

32. Hubble's law is used to determine the <u>distance</u> to galaxies.

33. The age of the universe is given by 1/H where H is <u>Hubble's</u> constant.

34. The fact that Hubble's constant appears to have changed with time supports the <u>Big Bang</u> theory.

35. The 3-K background radiation supports the <u>Big Bang</u> theory.

36. The oscillating universe theory predicts a <u>closed</u> universe.

37. Large stars have <u>shorter</u> lifetimes than small stars.

38. Pulsars are remanents of <u>supernova</u>.

Matching

(Choose the appropriate answer from the list on the right.)

<u>f</u> 1. cosmology

<u>i</u> 2. cosmos

<u>k</u> 3. Hubble's law

<u>h</u> 4. Big Bang theory

<u>m</u> 5. Steady state theory

<u>a</u> 6. 3-K background radiation

<u>r</u> 7. open universe

<u>l</u> 8. closed universe

<u>p</u> 9. red giant

<u>n</u> 10. white dwarf

<u>t</u> 11. black dwarf

<u>c</u> 12. supernova

<u>g</u> 13. pulsar

<u>o</u> 14. black hole

<u>d</u> 15. quasar

<u>s</u> 16. event horizon

<u>b</u> 17. Schwarzchild radius

<u>q</u> 18. oscillating universe

<u>j</u> 19. age of the universe

<u>e</u> 20. accretion disk

a. supports the Big Bang theory

b. the outer limit of the event horizon

c. the most energetic of all stellar explosions

d. shows a very large redshift

e. spiraling matter falling in-ward due to gravitational force

f. the study of the nature, structure, and origin of the universe

g. a rapidly rotating neutron star

h. expansion of the universe from a presumed primeval explosion

i. the world or universe re-garded as an orderly system

j. equals 1/H

k. H = v/d

l. two parallel lightbeams would eventually meet

m. assumes the continuous creation of matter

n. low-mass star with white hot surface

o. its velocity of escape is equal to or greater than the velocity of light

p. completing the hydrogen burning stage

(continued)

q. an expanding and contracting
   universe

r. two parallel light beams
   would eventually diverge

s. the surface of a black hole

t. a final state for a star

Solutions to Extended View Exercises

Introduction. Conversion Factors

1. 1 m/3 = 0.333 m = 3.33 cm

2. 30 mi (1.61 km/mi) = 48 km

3. 2 L (1.06 qt/L)(2 pt/qt) = 4.2 pt

4. 1000 kg (2.2 lb/kg) = 2,200 lb
    which is 200 lb greater than a 2000 lb ton.

5. (a) 22 lb (1 kg/2.2 lb) = 10 kg, etc.

6. 62 in. (2.54 cm/in.) = 157 cm = 1.57 m

    120 lb (1 kg/2.2 lb) = 54.5 kg

7. ¼ lb x 12 = 3 lb (1 kg/2.2 lb) = 1.4 kg

8. (a) 10 ft (0.305 m/ft) = 3.05 m,

    (b) 10 gal (4 qt/gal)(0.94 L/qt) = 37.6 L,

    (c) 1 mi = 1.61 km,

    (d) 2000 lb (1 kg/2.2 lb) = 909 kg

9. (a) 8 in. tall [likely]; (b) 300 mm = 0.30 m ≈ 1 ft
    [likely]; (c) 2 m (3.28 ft/m) = 6.6 ft [unlikely];
    (d) 65 kph = 40 mph [unlikely]; (e) 10 L =
    10.6 qt ≈ 21 pt ≈ 42 servings (8 oz) Pretty close.

10. 10 ft (0.305 m/ft) = 3.05 m, and 12 ft = 3.66 m.
    $A = l \times w = 3.05 \text{ m} \times 3.66 \text{ m} = 11.16 \text{ m}^2$.

11. 300 cubits (1.5 ft/cubit)(0.305 m/ft) ≈ 137 m, and
    50 cubits ≈ 23 m, and 30 cubits ≈ 14 m.

    $V = l \times w \times h$ = 44,114 cubic meters.

Extendend View Solutions

Chapter 1   Describing Motion

1.   A: $v = d/t = 100$ km$/2$ h $= 50$ km/h

  B: $v = d/t = 100$ km$/2.5$ h $= 40$ km/h

2.   Car: $v = d/t = 50$ km$/0.5$ h $= 100$ km/h

  Train: $v = d/t = 90$ km$/1$ h $= 90$ km/h

3.   (a) $d = vt = (0.50$ m/s$)(1$ s$) = 0.50$ m.   Similarly,
  (b) $d = 10$ m, (c) $d = 30$ m.

4. $d = vt = (300$ km/h$)(2$ h$) = 600$ km, east.

5. $d = vt = (2.0$ m/s$)(10$ s$) = 20$ m

6.   (a) $t = d/v = (160$ km$)/(160$ km/h$) = 1$ h. Similarly,
  (b) $t = 0.625$ h, (c) $t = 0.00625$ h.

7.   35 km $(0.62$ mi/km$) = 21.7$ mi, $t = 20$ min $= 1/3$ h.
  $v = d/t = 21.7$ mi$/(1/3$ h$) = 65.1$ mi/h

8. $d = \frac{1}{2}at^2 = \frac{1}{2}(4.0$ m/s$^2)(5.0$ s$)^2 = 50$ m

  $v = at = (4.0$ m/s$^2)(5.0$ s$) = 20$ m/s

9.   (b) $t = \sqrt{2d/g} = [2(1.22$ m$)/9.8$ m/s$^2]^{\frac{1}{2}} = 5.0$ s

10. $d_1 = \frac{1}{2}a_1t^2 = \frac{1}{2}(8.0$ m/s$^2)(5.0$ s$)^2 = 100$ m

  Similarly, $d_2 = 125$ m, and $d_2 - d_1 = 25$ m.

13.   (b) Area under curve: $(8.0$ m/s x 4.0 s$) +$
  $\frac{1}{2}(8.0$ m/s$)(5.0$ s$) = 32$ m $+ 20$ m $= 52$ m.

14. $d = v_o t + \frac{1}{2}at^2$

  $= (2.5$ m/s$)(8.0$ s$) + \frac{1}{2}(0.50$ m/s$^2)(8.0$ s$)^2 = 36$ m

15. $d = v_o t + \frac{1}{2}at^2$

  $= (8.0$ m/s$)(10$ s$) + \frac{1}{2}(4.0$ m/s$^2)(10$ s$)^2 = 280$ m

16. $t = 1$ s, $d = v_o t - \frac{1}{2}at^2$

  $= (12$ m/s$)(1$ s$) - \frac{1}{2}(8.0$ m/s$^2)(1$ s$)^2 = 8$ m.

  Simlarly for $t = 5$ s, $d = -40$ m (from $t = 0$
  position in direction opposite initial motion).

Chapter 2 The Conservation of Energy in Nonconservative Systems

1. $E = \frac{1}{2}mv^2 = \frac{1}{2}(0.15 \text{ kg})(7.6 \text{ m/s})^2 = 4.33 \text{ J}$

    $E_o = mgh = (0.15 \text{ kg})(9.8 \text{ m/s}^2)(3.0 \text{ m}) = 4.41 \text{ J}$

    $Q = E_o - E = 4.41 - 4.33 = 0.08 \text{ J}$

2. $mgh = mgh_o - Q$, and $h = h_o - Q/mg =$

    $0.20 - 0.050/(0.10)9.8) = 0.15 \text{ m}$

3. (a) $v = \sqrt{2gh_o} = [2(9.8)(0.20)]^{\frac{1}{2}} = 2.0 \text{ m/s}$

    (b) $v = \sqrt{2gh} = [2(9.8)(0.15)]^{\frac{1}{2}} = 1.7 \text{ m/s}$

4. $Q = mgh_o - \frac{1}{2}mv^2 = (30)(9.8)(10) - \frac{1}{2}(30)(5.0)^2$

    $= 2940 - 375 = 2526 \text{ J}$

5. (a) $v = (2gh_o)^{\frac{1}{2}} = [2(9.8)(1.5)]^{\frac{1}{2}} = 5.4 \text{ m/s}$

    (b) $Q = mg(h - h_o) = (0.50)(9.8)[1.5 - 1.2] = 1.5 \text{ J}$

    (c) $mgh/mgh_o = h/h_o = 1.2/1.5 = 0.80$

6. $E_o = mgh$ and 75 percent at bottom,

    $(0.75)mgh = \frac{1}{2}mv^2$. Then,

    $v = [(0.75)2gh]^{\frac{1}{2}} = [0.75)(=2(9.8)(5.0]^{\frac{1}{2}} = 8.6 \text{ m/s}$

8. $E = \frac{1}{2}mv^2 = \frac{1}{2}(850)(25)^{\frac{1}{2}} = 265,625 \text{ J} = 0.27 \text{ MJ}$

    $Eff = E/1.80 \text{ MJ} = 0.27/1.80 = 0.15$

Chapter 3 More on Collisions

1. $\Delta p = p_2 - p_1 = mv - (mv) = 2mv$

2. $3M(2 \text{ m/s}) = mv$, and $v = 6 \text{ m/s}$

3. $m_o v_o = (m_o + m)v$

    $v = m_o v_o/(m_o + m) = (80)(6)/100 = 4.8 \text{ m/s}$

4. $mv_o = 2 mv$, and $v = v_o/2 = 50/2 = 25 \text{ km/h}$

Extended View Solutions

(Chapter 3 continued)

5. $3m(90) = 3m(50) + mv$, and $v = 270 - 150 = 125$ km/h

6. $K/K_o = \frac{1}{3}m(13.9)^2 + \frac{1}{2}m(33.3)^2/\frac{1}{3}m(25)^2 = 0.90$

    so 0.10 or 10 percent loss.

7. $mv - 2m(v/2) = 0$

8. $2m(20) - m(10) = 3mv$, and $v = 10$ km/h

9. $m(20) - 3m(10) = 4mv$, and $v = -2.5$ km/h in the direction of the more massive car.

10. $m(20) + m(10) = 2mv$, and $v = 15$ km/h

Chapter 5 Inverse-square Laws

1. $F_2 = (r_1/r_2)^2 F_1 = (1/1.5)^2 F_1 = (0.44)F_1$

2   $F_2 = (r_1/r_2)^2 F_1 = (10/4)^2 = (6.25)F_1$

4. $g_2 = (r_1/r_2)^2 g_1 = (2)^2(9.8) = 39.2$ m/s^2

5. (a) $r_2 = (F_1/F_2)^{\frac{1}{2}} r_1 = (1/10)^{\frac{1}{2}} r_1 = r_1/3.2$

    (b) $F_2 = 0.9\ F_1$ (decreased by 1/10).

         $r_2 = (1/10)^{\frac{1}{2}} r_1 = (1.05)r_1$

6. (a) $r/R_e = (g_e/g)^{\frac{1}{2}} = (9.8/4.9)^{\frac{1}{2}} = 1.4$

    (b) $r/R_e = (9.8/29.4)^{\frac{1}{2}} = 0.58$

7. Masses would have to brought infinitely close together, which is physically impossible.

8. $r_2 = (F_1/F_2)^{\frac{1}{2}} r_1 = (500/200)^{\frac{1}{2}} = 1.6$

9. $g_2/g_1 = (Gm_2/r_2^2)/(Gm_1/r_1^2) = (m_2/m_1)(r_1/r_2)^2$

10. (a) $g_2 = (m_2/m_1)(r_1/r_2)^2 g_1 = (1/2)(1/3)^2 g_1 = g_1/18$

    (b) $g_2 = g_1/18 = 9.8/18 = 0.54$ m/s^2

11. $g_2 = (R_e/60\ R_e)^2 g_1 = (0.00028)(9.8) = 0.0027$ m/s^2

Chapter 8 Density and Volume

1. $\rho = m/V = 0.50 \text{ kg}/0.96 \text{ m}^3 = 0.52 \text{ kg/m}^3$

3. $D = w/V = mg/V = (\rho V)g/V = \rho g$

4. $V = 1.0 \times 0.80 \times 0.17 = 0.60 \text{ m}^3$

    (a) $m = \rho V = (1.29 \text{ kg/m}^3)(0.60 \text{ m}^3) = 0.77 \text{ kg}$

    (b) $0.77 \text{ kg } (2.2 \text{ lb/kg}) = 1.7 \text{ lb}$

5. $V_{Al} = (\rho_{Fe}/\rho_{Al})V_{Fe} = (7900/2700)V_{Fe} = (2.9)V_{Fe}$

6. $V_{ice} = (\rho_{w}/\rho_{ice})V_{w} = (1.0/0.92)V_{w} = (1.08)V_{w}$

7. $V = w/D = 100 \text{ lb}/(62.4 \text{ lb/ft}^3) = 1.6 \text{ ft}^3$

8. $V_{Br} = (\rho_{Au}/\rho_{Br})V_{Au}$
    $= (19.3/8.5)(1 \text{ cm}^3) = 2.27 \text{ cm}^3$

9. $2 \text{ gal } (0.13 \text{ ft}^3/\text{gal}) = 0.26 \text{ ft}^3$
    $w = DV = (62.4 \text{ lb/ft}^3)(0.26 \text{ ft}^3) = 16 \text{ lb}$

10. $V = 3.0 \text{ ft} \times 1.5 \text{ ft} \times 1.0 \text{ ft} = 4.5 \text{ ft}^3$
    $w_s - w_w = (D_s - D_w)V = (64.3 - 62.4)(4.5) = 8.6 \text{ lb}$

Chapter 11 Fahrenheit to Celsius and Back

1. (b) $T_F = (9/5)T_C + 32 = (9/5)(30) + 32 = 86°F$
    (c) $T_F = (9/5)(-10) + 32 = 14°F$

2. (a) $T_C = (5/9)(T_F - 32) = (5/9)(0 - 32) = 18°C$
    (b) $T_C = (5/9)(114 - 32) = 46°C$
    (c) $T_C = (5/9)(-13 - 32) = -25°C$

3. $T_F = (9/5)T_C + 32 = (9/5)(15) + 32 = 59°F$

4. (a) $T_C = (5/9)(T_F - 32) = (5/9)(77 - 32) = 25°C$
    (b) $T_C = (5/9)(59 - 32) = 15°C$

6. $T_F = (9/5)T_C + 32 = (9/5)(40) + 32 = 104°F$

(Chapter 11 continued)

7. $T_K = T_C + 273 = -40 + 273 = 233$ K

8. $T_C = T_K - 273 = 288 - 273 = 15^{\circ}C = 59^{\circ}F$ (cf. Ex. 3)

9. $T_C = (5/9)(T_F - 32) = (5/9)(104 - 32)$

$$= 40^{\circ}C \text{ (cf. Ex.6)}$$

$T_K = T_C + 273 = 40 + 273 = 313$ K

Chapter 13   Thermodynamics, Heat Engines and Heat Pumps

1.   (a)   2 power strokes/rev (3000 rev/min) = 6000/min

   (b)   More cylinders, more power strokes/rev.

2.   (c) $T_{cold} = 5^{\circ}C + 273 = 278$ K

   $T_{hot} = 25^{\circ}C + 273 = 298$ K

   Eff $= 1 - (T_{cold}/T_{hot}) = 1 - (278/298) = 0.067$
   $(= 6.7 \text{ percent})$

3.   (a) $T_{hot} = 540^{\circ}C + 273 = 813$ K

   $T_{cold} = 20^{\circ}C + 273 = 293$ K

   Eff $= 1 - (T_{cold}/T_{hot}) = 1 - (293/813) = 0.64$
   $(= 64 \text{ percent})$

4. $T_{hot} = 250^{\circ}C + 273 = 523$ K

   $T_{cold} = 60^{\circ}C + 273 = 333$ K

   $Eff_1 = 1 - (T_{cold}/T_{hot}) = 1 - (333/523) = 0.36$

   $Eff_2 = 2\ Eff_1 = 2(0.36) = 0.72 = 1 - (333/T_{hot})$

   $(T_{hot})_2 = 333/(1 - 0.72) = 1189$ K $(- 273 = 916^{\circ}C)$

5.   (a) cop $= T_{cold}/(T_{hot} - T_{cold})$
   $= 1/[(T_{hot}/T_{cold}) - 1]$.   So, cop >1 when
   $T_{hot}/T_{cold} < 2$ or $T_{hot} < 2\ T_{cold}$, which is the
   general case for refrigerators.

   (b)   cop $= Q_{out}/W_{in} = 2.5$, and $Q_{out} = (2.5)W_{in}$

(Chapter 13 continued)

    (c)   cop $= 3.0 = Q_{out}/W_{in} = Q_{out}/4000$ J

         and $Q_{out} = (3.0)(4000$ J$) = 12,000$ J

           $Q_{in} = Q_{out} - W = 12,000 - 4000 = 8000$ J

    (d)   As $(T_{hot} - T_{cold})$ increases, cop decreases.

Chapter 14   Vibrations and Waves

1. $T = 1/f = 1/10 = 0.10$ s

2.   (a) $T =$ s/cycle $= 60/2 = 30$ s

       (b) $f = 1/T = 1/30$ Hz

3.   (a) $f =$ cycles/t or cycles $= ft$.

       $ft = 5 \times 2 = 10$ cycles, and $ft = 5 \times 0.6 = 3$ cycles

       (b) $T = 1/f = 1/5 = 0.20$ s

4.   kHz $\approx 10^3$ Hz and MHz $\approx 10^6$ Hz

      $T = 1/f \approx 1/10^3 = 10^{-3}$ s and $T \approx 1/10^6 = 10^{-6}$ s

5. $v = \lambda/t = 2/2 = 1$ m/s

6.   (a) $T =$ sec/cycle $= 3/4$ s, and $f = 1/T = 4/3$ Hz

7. $\lambda = c/f \approx 10^8/10^{14} = 10^{-6}$ m

8.   (a) $I = (r_o/r)^2 I_o = (1/4)^2 I_o = I_o/16$

       (b) $I = (3)^2 I_o = 9I_o$. $(r = r_o/3)$

9. $T = 1/f = 1/(1/3) = 3$ s

Chapter 15   Sound and Music

1. $c/v_s \approx (300$ million m/s$)/(300$ m/s$) = 1$ million

2. $\Delta v_s = 0.6(T_2 - T_1) = 0.6(25^o - 0^o) = 15$ m/s

3.   (a) $\Delta T = \Delta v_s/0.6 = (6$ m/s$)/0.6 = 10^o$C

       (b) $\Delta T = (10$ m/s$)/0.6 = 16.7^o$C

4. $f = v_s/\lambda = 340/2 = 170$ Hz

Extended View Solutions

(Chapter 15 continued)

5. $\lambda_1 = v_s/f_1 = 340/20$ Hz $= 17$ m

       $\lambda_2 = 340/20,000$ Hz $= 0.017$ m

6. $v = Mv_s = (1.5)(330$ m/s$) = 495$ m/s

Chapter 17   Series and Parallel Resistances

1.   (a) $R_s = R_1 + R_2 = 4 + 6 = 10$

       (b) $R_p = R_1R_2/(R_1 + R_2) = (4)(6)/(4 + 6) = 2.4$ Ω

2. $1/R_p = 1/R_1 + 1/R_2 = (R_1 + R_2)/R_1R_2$, and invert.

3.   (a) $R_s = R_1 + R_2 + R_3 = 1 + 2 + 3 = 6$ Ω

       (b) $1/R_p = 1/R_1 + 1/R_2 + 1/R_3 = 1/1 + 1/2 + 1/3$

       $= 6/6 + 3/6 + 2/6 = 11/6$, and $R_p = 6/11\Omega = 0.55$ Ω

4.   (Single) R_1, R_2, R_3; (Series) R_1R_2, R_1R_3, R_2R_3,

       $R_1R_2R_3$; (Parallel) same 4 combinations as series;

       (Series-parallel) R_1-R_2R_3, R_2-R_1R_3, R_3-R_1R_2

5.   (a) $R_{p1} = (2)(4)/(2 + 4) = 8/6$ Ω

             $R_{p2} = (1)(3)/(1 + 3) = 3/4$ Ω

             $R_s = R_{p1} + R_{p2} = 8/6 + 3/4$ Ω

                         $= 16/12 + 9/12 = 25/12 = 2.1$ Ω

       (b) $I = V/R_s = 12/2.1 = 5.7$ A

6.   (a)   100 W;   (b) $I = P/V = 100/120 = 0.83$ A;

       (c) $R = V/I = 120/0.83 = 144$ Ω

7. $R_s = 100 + 100 = 200$ Ω.

       (a) $I = V/R_s = 120/200 = 0.60$ A

       (b) $P = IV = (0.60)(120) = 72$ W

       (c)   Currrent divides equally, 0.30 A

8. $R_p = R_1R_2/(R_1 + R_2) = (5)(20)/(5 + 20) = 4\,\Omega$.

   (a) $I = V/R_p = 12/4 = 3.0$ A

   (b) $P = IV = (3.0)(12) = 36$ W

   (c) $I_1 = V/R_1 = 12/5 = 2.4$ A; $I_2 = 12/20 = 0.60$ A

   (d) $P_1 = I_1V = (2.4)(12) = 28.8$ W,

      $P_2 = I_2V = (0.60)(12) = 7.2$ W

9.   (a) $P = IV = (1.4)(120) = 168$ W

   (b) $R = V/I = 120/1.4 = 86\,\Omega$

10. $W = Pt = (0.10$ kW$)(2$ h/day$) = 0.20$ kWh/day.

   Cost $= 0.20$ kWh/day $(30$ days$)($\$0.08$/kWh$) = \$0.48$.